Annals of Mathematics Studies

Number 89

INVARIANT FORMS ON GRASSMANN MANIFOLDS

BY

WILHELM STOLL

PRINCETON UNIVERSITY PRESS
AND
UNIVERSITY OF TOKYO PRESS

PRINCETON, NEW JERSEY
1977

Published in Japan exclusively by
University of Tokyo Press;
In other parts of the world by
Princeton University Press

Printed in the United States of America
by Princeton University Press, Princeton, New Jersey

Library of Congress Cataloging in Publication data will
be found on the last printed page of this book

CONTENTS

PREFACE vii

GERMAN LETTERS ix

INTRODUCTION 3

1. FLAG SPACES 11

2. SCHUBERT VARIETIES 27

3. CHERN FORMS 35

4. THE THEOREM OF BOTT AND CHERN 43

5. THE POINCARÉ DUAL OF A SCHUBERT VARIETY 57

6. MATSUSHIMA'S THEOREM 64

7. THE THEOREMS OF PIERI AND GIAMBELLI 82

APPENDIX 103

REFERENCES 110

INDEX 113

PREFACE

Schubert varieties describe the cohomology of Grassmann manifolds. The Poincaré duals of the Schubert varieties generate the vector space of invariant forms, which as an exterior algebra is isomorphic to the cohomology ring of the Grassmann manifold. In fact, Giambelli's theorem asserts that this algebra is generated by the basic Chern forms. Thus the theory of invariant forms on Grassmann manifolds is important for the study of vector bundles and characteristic classes.

I became interested in this subject matter because of its applications to value distribution theory. Bott and Chern studied the equidistribution of the zeroes of holomorphic sections in holomorphic vector bundles. Recently, Cowen considered Schubert zeroes of holomorphic vector bundles on pseudoconcave spaces. In order to extend this theory to pseudoconvex spaces, a "deficit term," which requires the calculation of a certain invariant form on a Grassmann manifold, has to be identified. On this matter I consulted Y. Matsushima who obtained a very useful theorem. His proof is long, difficult and uses deep results about Lie algebras. Later I found a simpler proof which uses only fiber integration and elementary considerations. I employ the same method to obtain a number of known results such as the duality theorem, the theorem of Pieri and the representation theorem of Bott and Chern. Pieri's theorem easily implies Giambelli's theorem. The proof of Matsushima's theorem given here is about as difficult as the proof of the duality theorem. It uses fiber integration in the double intersection diagram. The theorem of Pieri is much more difficult to obtain and requires fiber integration in a triple intersection diagram. Fiber integration along smooth fibers is well known, but does not suffice here. The extension to fibers with singularities is not trivial, but has been established in Ch. Tung's thesis.

After this monograph was written, but before it was sent to the publisher, Matsushima received a latter from J. Damon in which Damon proves Matsushima's theorem based on Damon [9] and [10]. This proof uses the Gysin homomorphism computed by a residue calculus.

Most of the results in this monograph are known. The method of proof is new, especially in the case of Matsushima's theorem. The topic is not easily accessible. Therefore an introduction has been written here in order to provide a clear, coherent, intransically formulated account which will be useful for applications to value distribution theory, and which may have a wider appeal as well.

I thank the National Science Foundation for partially supporting this research under Grant MPS 75-07086.

WILHELM STOLL

GERMAN LETTERS

A	𝔄	a	𝔞
B	𝔅	b	𝔟
C	ℭ	c	𝔠
D	𝔇	d	𝔡
E	𝔈	e	𝔢
F	𝔉	f	𝔣
G	𝔊	g	𝔤
H	ℌ	h	𝔥
I	ℑ	i	𝔦
J	𝔍	j	𝔧
K	𝔎	k	𝔨
L	𝔏	l	𝔩
M	𝔐	m	𝔪
N	𝔑	n	𝔫
O	𝔒	o	𝔬
P	𝔓	p	𝔭
Q	𝔔	q	𝔮
R	ℜ	r	𝔯
S	𝔖	s	𝔰
T	𝔗	t	𝔱
U	𝔘	u	𝔲
V	𝔙	v	𝔳
W	𝔚	w	𝔴
X	𝔛	x	𝔵
Y	𝔜	y	𝔶
Z	ℨ	z	𝔷

Invariant Forms on Grassmann Manifolds

INTRODUCTION

Main results

The Grassmann manifold of all (p+1)-dimensional linear subspaces of the (n+1)-dimensional complex vector space V has dimension $d(p,n) = (p+1)(n-p)$. If $x \in G_p(V)$, let $E(x)$ be the (p+1)-dimensional linear subspace of V defined by x and let $\ddot{E}(x)$ be the p-dimensional projective plane defined by x and contained in the projective space $\mathbf{P}(V) = G_0(V)$. As an algebraic subvariety of $\mathbf{P}(\underset{p+1}{\wedge} V)$, the degree of $G_p(V)$ is denoted by $D(p,n)$. A hermitian metric $\mathfrak{l}$ is selected on V, which determines the unitary group $\mathfrak{U} = \mathfrak{U}(V,\mathfrak{l})$.

Let $\mathfrak{S}(p,n)$ be the set of all "symbols" $\mathfrak{a} = (a_0,\cdots,a_p)$ where $a_0,\cdots,a_p$ are integers with $o \leq a_0 \leq \cdots \leq a_p \leq n-p$. If $a_q \leq b_q$ for $q = 0,1,\cdots,p$ write $\mathfrak{a} \leq \mathfrak{b}$. Define $\vec{\mathfrak{a}} = a_0 + \cdots + a_p$. The flag manifold $F(\mathfrak{a})$ consists of all $v = (v_0,\cdots,v_p)$ with $v_q \in G_{a_q+q}(V)$ for $q = 0,1,\cdots,p$ such that

$$E(v_0) \subset E(v_1) \subset \cdots \subset E(v_p) .$$

Then $F(\mathfrak{a})$ is a compact, connected, complex manifold of dimension $d(\mathfrak{a})$, and $\mathfrak{U}$ acts transitively on $F(\mathfrak{a})$. Take $v \in F(\mathfrak{a})$. The Schubert variety $S(v,\mathfrak{a})$ consists of all $x \in G_p(V)$ such that $\dim E(x) \cap E(v_q) \geq q+1$ for all $q = 0,\cdots,p$. Then $S(v,\mathfrak{a})$ is an irreducible analytic subset of dimension $\vec{\mathfrak{a}}$ of $G_p(V)$. The Schubert family

$$S(\mathfrak{a}) = \{(x,v) \in G_p(V) \times F(\mathfrak{a}) \mid x \in S(v,\mathfrak{a})\}$$

is an irreducible, analytic subset of $G_p(V) \times F(\mathfrak{a})$. The projections $\pi : S(\mathfrak{a}) \to G_p(V)$ and $\sigma : S(\mathfrak{a}) \to F(\mathfrak{a})$ are locally trivial and surjective

(Cowen [8] and Lemma 2.1). Hence the fiber integration operators π_* and σ_* are defined (Tung [33]). There is one and only one volume form $\Omega_{\mathfrak{a}} > 0$ on $F(\mathfrak{a})$ such that $\Omega_{\mathfrak{a}}$ is invariant under the action of $\mathfrak{U}$ and

$$\int_{F(\mathfrak{a})} \Omega_{\mathfrak{a}} = 1 .$$

Then $c(\mathfrak{a}) = \pi_* \sigma^*(\Omega_{\mathfrak{a}}) \geq 0$ is a form of bidegree (a,a) on $G_p(V)$ with $a = d(p,n) - \vec{\mathfrak{a}}$. The form $c(\mathfrak{a})$ is invariant under the action of $\mathfrak{U}$ and is the Poincaré dual of $S(v, \mathfrak{a})$ for each $v \in F(\mathfrak{a})$. The forms $c(\mathfrak{a})$ generate the cohomology of $G_p(V)$.

Take $\mathfrak{a} \in \mathfrak{S}(p,n)$. Let r be an integer with $o \leq r \leq d(p,n) - \vec{\mathfrak{a}}$. Assume that $s = d(\mathfrak{a}) - r \geq 0$. Define

$$\Delta(\mathfrak{a},r) = \{ \mathfrak{b} \in \mathfrak{S}(p,n) \mid \mathfrak{b} \geq \mathfrak{a} \text{ and } \vec{\mathfrak{b}} = \vec{\mathfrak{a}} + r\} .$$

Let λ be a form of class C^∞ and degree $2s$ on $F(\mathfrak{a})$. Assume λ is invariant under the action of $\mathfrak{U}$. Define $\Lambda = \pi_* \sigma^*(\lambda)$. *Matsushima's Theorem* [21] asserts the existence of constants $\gamma_{\mathfrak{b}}$ such that

$$\Lambda = \sum_{\mathfrak{b} \in \Delta(\mathfrak{a},r)} \gamma_{\mathfrak{b}}\, c(\mathfrak{b})$$

(Theorem 6.12). Clearly, Λ is a linear combination of $c(\mathfrak{b})$. The point is that $\gamma_{\mathfrak{b}} = 0$ outside $\Delta(\mathfrak{a},r)$. The proof given here uses the wedge product formula of fiber integration [22] and the double intersection diagram.

The tautological bundle

$$S_p(V) = \{(x, \mathfrak{v}) \in G_p(V) \times V \mid \mathfrak{v} \in E(x)\}$$

is a subbundle of the trivial bundle $G_p(V) \times V$. Let $Q_p(V)$ be the quotient bundle. The hermitian metric $\mathfrak{l}$ on V defines a hermitian metric along the fibers of $G_p(V) \times V$ and $S_p(V)$. The bundle $S_p(V)^\perp$

orthogonal to $S_p(V)$ is isomorphic to $Q_p(V)$. A hermitian metric is defined along the fibers of $Q_p(V)$. The associated Chern forms $c_0[p], \cdots, c_{n-p}[p]$ of $Q_p(V)$ are invariant under the action of $\mathfrak{U}$. Define $c_q[p] = 0$ if $q < 0$ or $q > n-p$. The *Theorem of Giambelli* [13] asserts that

$$c(\mathfrak{a}) = \det c_{n-p-a_i-i+j}[p]$$

(Theorem 7.5), and is based on the Theorem of Pieri [21], which computes $c(\mathfrak{a}) \wedge c_h[p]$ in terms $c(\mathfrak{b})$ (Chern [5], Hodge [18], and Vesentini [34]). The proof given here (Theorem 7.4) uses fiber integration in the triple intersection diagram. Specific care is taken (Lemma 6.7) to clarify the general position mentioned by Chern [5].

Let p, q and μ be integers with $o \leq q \leq p \leq \mu \leq n$. Define

$$F_{pq} = \{(x,y) \in G_q(V) \times G_p(V) \mid E(x) \subseteq E(y)\} .$$

Let $\tau : F_{pq} \to G_p(V)$ and $\pi : F_{pq} \to G_q(V)$ be the projections. The Bott-Chern Representation Theorem [3] states that

$$D(q-1,p-1)\, c_{\mu-p}[p] = \tau_* \pi^*(c_{\mu-q}[q] \wedge c_1[q]^{(p-q)q}) .$$

The proof announced in [28] will be finally given here.

This monograph is almost self-contained. Some elementary facts on Grassmann manifolds and flag manifolds are not proved here. For a proof that the open Schubert cell subdivision of $G_p(V)$ defines a CW-complex see Milnor-Stasheff [23]. The proof that $S(v, \mathfrak{a})$ is irreducible, and that the open Schubert cell $S^*\langle v, \mathfrak{a}\rangle$ is biholomorphically equivalent to $\mathbb{C}^{\vec{\mathfrak{a}}}$ follows Chern [7] and is given in the appendix for the convenience of the reader.

Application to value distribution

This paper was written with applications to value distribution in mind. Such an application shall be outlined here. For details see [32].

Let W be a holomorphic vector bundle of fiber dimension k over an irreducible compact complex space N. The vector space V of global holomorphic sections has finite dimension $n+1$. Assume that $p = n-k \geq 0$. The evaluation map $e : N \times V \to W$ is defined by $e(x,s) = s(x)$. Assume that e is surjective. Then W is said to be ample. If $x \in N$, define $e_x : V \to W_x$ by $e_x(s) = s(x)$. Take $\alpha = (a_0, \cdots, a_p) \in \mathfrak{S}(p,n)$ and $v = (v_0, \cdots, v_p) \in F(\alpha)$. The *Schubert zero set*

$$S_W(v, \alpha) = \bigcap_{q=0}^{p} \{x \in N \mid \dim e_x(E(v_q)) \leq a_q\}$$

is analytic in N.

Let M be an irreducible complex space of dimension m. Assume a holomorphic map $f : M \to N$ is given. When is $f(M) \cap S_W(v, \alpha) \neq \emptyset$ for many $v \in F(\alpha)$? The family $S(\alpha)$ is admissible in the sense of Tung [33]. Therefore a solution can be given along the following lines.

Take a hermitian metric $\mathfrak{l}$ on V. By $e : N \times V \to W$, a quotient hermitian metric $\mathfrak{l}$ is defined along the fibers of W. Let $c_q(W, \mathfrak{l}) \geq 0$ be the associated Chern forms of W for $q = 0, \cdots, k$. Define $c_q(W, \mathfrak{l}) = 0$ if $q < 0$ or if $q > k$. Take $\alpha \in \mathfrak{S}(p,n)$. Define a form of class C^∞ and bidegree (s,s) with $s = d(p,n) - \vec{\alpha}$ by

$$c_W(\alpha) = \det c_{k-a_i-i+j}(W, \mathfrak{l})$$

on N. Then $c_W(\alpha) \geq 0$ is closed.

Assume that M carries *a pseudoconvex exhaustion* τ. This means, that a non-negative function τ of class C^∞ is given on M such that $dd^c\tau \geq 0$ and such that

$$M[r] = \{x \in M \mid \tau(x) \leq r\}$$

is compact for each $r \geq 0$. Also define

$$M(r) = \{x \in M \mid \tau(x) < r\} \qquad M\langle r\rangle = \{x \in M \mid \tau(x) = r\}.$$

Define $\upsilon = dd^c\tau$. Assume that $q = m + \vec{\alpha} - d(p,n) \geq 0$. If $q > 0$, assume that $\upsilon > 0$ on some non-empty open subset of M. Define

$$A_f(r,\mathfrak{a}) = \int_{M[r]} f^*(c_W(\mathfrak{a})) \wedge \upsilon^q \geq 0$$

$$T_f(r,\mathfrak{a}) = \int_0^r A_f(t,\mathfrak{a})dt \geq 0 \,.$$

Here $T_f(r,\mathfrak{a})$ is called the characteristic of f for $\mathfrak{a} \in \mathfrak{S}(p,n)$. Assume that there exists at least one $v \in F(\mathfrak{a})$ for which there exists at least one point $x \in M$, such that $f^{-1}(S_W(v,\mathfrak{a}))$ has dimension q at x. Then $T_f(r,\mathfrak{a}) \to \infty$ for $r \to \infty$. In [32], the following theorem is proved: If

$$\frac{A_f(r,\mathfrak{b})}{T_f(r,\mathfrak{a})} \to 0 \qquad \text{for } r \to \infty$$

for each $\mathfrak{b} \in \mathfrak{S}(p,n)$ with $\mathfrak{b} \geq \mathfrak{a}$ and with $\vec{\mathfrak{b}} = \vec{\mathfrak{a}} + 1$, then $f(M) \cap S_W(v,\mathfrak{a}) \neq \emptyset$ for almost all $v \in F(\mathfrak{a})$.

The proof is of interest here, since it involves invariant forms on Grassmann manifolds and the Theorems of Giambelli and Matsushima. Hence a short outline of the proof shall be given. For details see [32].

Take $x \in N$. The kernel S_x of $e_x : V \to W_x$ is a $(p+1)$-dimensional linear subspace of V. One and only one $\phi(x) \in G_p(V)$ exists such that $E(\phi(x)) = S_x$. The map $\phi : N \to G_p(V)$ is holomorphic with $c_W(\mathfrak{a}) = \phi^*(c(\mathfrak{a}))$ and $S_W(v,\mathfrak{a}) = \phi^{-1}(S(v,\mathfrak{a}))$. Here Giambelli's Theorem is used. According to Hirschfelder [14], [15] and Wu [35], a proximity form $\lambda_v \geq 0$ on $F(\mathfrak{a}) - \{v\}$ exists for each $v \in F(\mathfrak{a})$ such that $dd^c\lambda_v = \Omega_\mathfrak{a}$ on $F(\mathfrak{a}) - \{v\}$ and such that $g^*(\lambda_{g(v)}) = \lambda_v$ for all $g \in \mathfrak{U}$. On $G_p(V) - S(v,\mathfrak{a})$, define $\Lambda_v = \pi_*\sigma^*(\lambda_v) \geq 0$. Then $dd^c\Lambda_v = c(\mathfrak{a})$ on $G_p(V) - S(v,\mathfrak{a})$. Define $\Lambda_{W,v} = \phi^{-1}(\Lambda_v) \geq 0$ on $N - S_W(v,\mathfrak{a})$ with $dd^c\Lambda_{W,v} = c_W(\mathfrak{a})$. Take $v \in F(\mathfrak{a})$ and assume that $A = f^{-1}(S_W(v,\mathfrak{a}))$ is either empty or pure q-dimensional, which is the case for almost all $v \in F(\mathfrak{a})$. For each $z \in A$, a multiplicity $\theta_f^v(z) > 0$ is assigned. The *counting* *function*, the *valence function* and the *deficit* of f for $(v, \mathfrak{a})$ are defined for all $r > 0$ by

$$n_f(r,v) = \int_{A \cap M[r]} \theta_f^v \upsilon^q \geq 0$$

$$N_f(r,v) = \int_0^r n_f(t,v)\,dt \geq 0$$

$$D_f(r,v) = \int_{M[r]} f^*(\Lambda_{W,v}) \wedge \upsilon^{q+1} \geq 0 .$$

For almost all $r > 0$, the *compensation* function is defined by

$$m_f(r,v) = \int_{M\langle r\rangle} f^*(\Lambda_{W,v}) \wedge d^c\tau \wedge \upsilon^q \geq 0 .$$

The First Main Theorem

$$T_f(r, \mathfrak{a}) = N_f(r,v) + m_f(r,v) - D_f(r,v)$$

holds and extends the definition of $m_f(r,v)$ to all $r > 0$.

The integral average

$$\hat{\lambda}(x) = \int_{F(\mathfrak{a})} \lambda_v(x)\Omega_{\mathfrak{a}}(v) \geq 0$$

exists and is a form on $F(\mathfrak{a})$ invariant under the action of $\mathfrak{U}$. The form $\hat{\Lambda} = \pi_* \sigma^*(\hat{\lambda}) \geq 0$ is invariant under the action of $\mathfrak{U}$ on $G_p(V)$ and has bidegree (s,s) with $s = d(p,n) - \vec{\mathfrak{a}}$. Define $\Delta(\mathfrak{a}) = \Delta(\mathfrak{a},1)$. By Matsushima's Theorem constants $\gamma_{\mathfrak{a}\,\mathfrak{b}} \geq 0$ exist such that

$$\hat{\Lambda} = \sum_{\mathfrak{b} \in \Delta(\mathfrak{a})} \gamma_{\mathfrak{a}\mathfrak{b}} \, c(\mathfrak{a}) \, .$$

Define $\hat{\Lambda}_W = \phi^*(\hat{\Lambda})$. An exchange of integration shows that

$$\hat{m}_f(r, \mathfrak{a}) = \int_{F(\mathfrak{a})} m_f(r,v)\Omega_{\mathfrak{a}} = \int_{M\langle r \rangle} f^*(\hat{\Lambda}_W) \wedge d^c\tau \wedge \upsilon^q$$

$$\hat{D}_f(r, \mathfrak{a}) = \int_{F(\mathfrak{a})} D_f(r,v)\Omega_{\mathfrak{a}} = \int_{M[r]} f^*(\hat{\Lambda}_W) \wedge \upsilon^{q+1}$$

$$= \sum_{\mathfrak{b} \in \Delta(\mathfrak{a})} \gamma_{\mathfrak{a}\mathfrak{b}} \, A_f(r, \mathfrak{b}) \, .$$

Stokes' Theorem implies that $\hat{D}_f(r, \mathfrak{a}) = \hat{m}_f(r, \mathfrak{a})$. Therefore

$$T_f(r, \mathfrak{a}) = \int_{F(\mathfrak{a})} N_f(r,v)\Omega_{\mathfrak{a}} \, .$$

The set $B = \{v \in F(\mathfrak{a}) \mid f(M) \cap S_W(v, \mathfrak{a}) \neq \emptyset\}$ is measurable. Then

$$0 \leq b_f(\mathfrak{a}) = \int_B \Omega_{\mathfrak{a}} \leq 1 \, .$$

Because $N_f(r,v) = 0$ if $v \in F(\mathfrak{a}) - B$, integration over B and the inequality $N_f(r,v) \leq T_f(r, \mathfrak{a}) + D_f(r,v)$ imply that

$$0 \leq 1 - b_f(\mathfrak{a}) \leq \frac{\hat{D}_f(r,\mathfrak{a})}{T_f(r, \mathfrak{a})} = \sum_{\mathfrak{b} \in \Delta(\mathfrak{a})} \gamma_{\mathfrak{a}\mathfrak{b}} \frac{A_f(r, \mathfrak{b})}{T_f(r, \mathfrak{a})} \, .$$

Therefore if $A_f(r, \mathfrak{b})/T_f(r, \mathfrak{a}) \to 0$ for $r \to \infty$ for each $\mathfrak{b} \in \Delta(\mathfrak{a})$, then $b_f(\mathfrak{a}) = 1$, which means that $f(M) \cap S_W(v, \mathfrak{a}) \neq \emptyset$ for almost all $v \in F(\mathfrak{a})$. If $E(v_0) \subset \cdots \subset E(v_n)$ is a maximal flag, select a subflag $v_{\mathfrak{a}} = (v_{a_0}, v_{a_1+1}, \ldots, v_{a_p+p}) \in F(\mathfrak{a})$ for each $\mathfrak{a} \in \mathfrak{S}(p,n)$. Given $\mathfrak{a} \in \mathfrak{S}(p,n)$, then $\mathfrak{b} \in \mathfrak{S}(p,n)$ belongs to $\Delta(\mathfrak{a})$ if and only if $S(v_{\mathfrak{a}}, \mathfrak{a}) \subset S(v_{\mathfrak{b}}, \mathfrak{a})$ and $\vec{\mathfrak{b}} = \vec{\mathfrak{a}} + 1$. (See (2.8).)

1. FLAG SPACES

a) *Grassmann manifolds*

For any partially ordered set A denote

$$A[a,b] = \{x \in A \mid a \leq x \leq b\} \qquad A(a,b) = \{x \in A \mid a < x < b\}$$
$$A(a,b] = \{x \in A \mid a < x \leq b\} \qquad A[a,b) = \{x \in A \mid a \leq x < b\} .$$

In fact, a,b may not belong to A, but to a larger set. For instance $\mathbf{R}^+ = \mathbf{R}(0,+\infty)$ or $\mathbf{R}_+ = \mathbf{R}[0,+\infty)$ or $\mathbf{Z}[1,\sqrt{2}] = \{1\}$.

Let V be a complex vector space of dimension $n+1 > 0$. For $0 \neq \mathfrak{x} \in V$ define $\mathbf{P}(\mathfrak{x}) = \mathbf{C}\mathfrak{x}$. If $A \subseteq V$, define $\mathbf{P}(A) = \{\mathbf{P}(\mathfrak{x}) \mid 0 \neq \mathfrak{x} \in A\}$. Then $\mathbf{P}(V)$ is a connected, compact, complex manifold of dimension n called the *complex projective space* of V. The map $\mathbf{P} : V - \{0\} \to \mathbf{P}(V)$ is holomorphic and denoted by the same letter $\mathbf{P}$ for all vector spaces. Take $p \in \mathbf{Z}[0,n]$. The *Grassmann cone* of order p to V is defined by

$$\tilde{G}_p(V) = \{\mathfrak{x}_0 \wedge \cdots \wedge \mathfrak{x}_p \mid \mathfrak{x}_\mu \in V \text{ for } \mu = 0,\cdots,p\} \subseteq \underset{p+1}{\wedge} V .$$

The *Grassmann manifold* $G_p(V) = \mathbf{P}(\tilde{G}_p(V))$ of order p to V is a compact, connected, smooth complex submanifold of $\mathbf{P}(\underset{p+1}{\wedge} V)$ of dimension $d(p,n) = (n-p)(p+1)$ and of degree

$$D(p,n) = d(p,n)! \prod_{q=0}^{p} \frac{q!}{(n-q)!} .$$

(For instance, see [23] Proposition 2.7 for a proof.) If $p = 0$, then $G_0(V) = \mathbf{P}(V)$. If $p = n$, then $G_n(V) = \{\infty\}$ consists of exactly one point, denoted by ∞.

Take $x \in G_p(V)$. Then $x = \mathbb{P}(\mathfrak{x}_0 \wedge \cdots \wedge \mathfrak{x}_p)$ and

$$E(x) = \mathbb{C}\mathfrak{x}_0 + \cdots + \mathbb{C}\mathfrak{x}_p = \{\mathfrak{z} \in V \mid \mathfrak{z} \wedge \mathfrak{x}_0 \wedge \cdots \wedge \mathfrak{x}_p = 0\}$$

is a (p+1)-dimensional linear subspace of V, depending only on x. Therefore $G_p(V)$ parameterizes the set of (p+1)-dimensional, linear subspaces of $G_p(V)$ bijectively. The projective space $\ddot{E}(x) = \mathbb{P}(E(x))$ is a smooth submanifold of $\mathbb{P}(V)$ and is called a *p-plane* in $\mathbb{P}(V)$.

Let W be another complex vector space. The set $L(V,W)$ of all linear maps $V \to W$ is a complex vector space of $\dim V \cdot \dim W$. Abbreviate $L_V = L(V,V)$ and $L(x,y) = L(E(x),E(y))$ if $x \in G_p(V)$ and $y \in G_q(V)$. Obviously, $V^* = L(V,\mathbb{C})$ is the dual space of V. The open, dense subset $GL(V) = \{g \in L_V \mid g \text{ isomorphic}\}$ is a complex Lie group under composition, called the *general linear group* of V. Here $GL(V)$ acts holomorphically on V, V^*, $\underset{p}{\wedge} V$, $\underset{p}{\oplus} V$, $\underset{p}{\otimes} V$, $\mathbb{P}(V)$ and $G_p(V)$ as a group of biholomorphic maps. If $g \in GL(V)$ and if $x = \mathbb{P}(\mathfrak{x}_0 \wedge \cdots \wedge \mathfrak{x}_p) \in G_p(V)$, then

$$g(x) = \mathbb{P}(g(\mathfrak{x}_0) \wedge \cdots \wedge g(\mathfrak{x}_p))$$

$$g(E(x)) = E(g(x)) \qquad g(\ddot{E}(x)) = \ddot{E}(g(x)) .$$

Abbreviate $GL(x) = GL(E(x))$.

Points $x \in G_p(V)$ and $y \in G_q(V)$ are said to be *independent* if $E(x) \cap E(y) = 0$. If so, then $x \wedge y \in G_{p+q+1}(V)$ is defined by $E(x \wedge y) = E(x) \oplus E(y)$. Let $j_x : E(x) \to E(x) \oplus E(y)$ and $j_y : E(y) \to E(x) \oplus E(y)$ be the inclusions and $\pi_x : E(x) \oplus E(y) \to E(x)$ and $\pi_y : E(x) \oplus E(y) \to E(y)$ the projections. Then $\pi_x \circ j_x$ and $\pi_y \circ j_y$ and $j_x \circ \pi_x + j_y \circ \pi_y$ are the identities. Moreover $\pi_y \circ j_x = 0$ and $\pi_x \circ j_y = 0$. If $q = n-p-1$, then $E(x) \oplus E(y) = V$.

Take $e \in G_p(V)$. The isotropy group $P_e = \{g \in GL(V) \mid g(e) = e\}$ is a closed subgroup of $GL(V)$. Let $\rho(g) = g P_e$ be the left coset of $g \in GL(V)$ and let $GL(V)/P_e$ be the left coset space. A holomorphic fiber bundle $\pi : GL(V) \to G_p(V)$ is defined by $\pi(g) = g(e)$. One and only one map $\iota: GL(V)/P_e \to G_p(V)$ exists such that $\iota \circ \rho = \pi$. Here ι is bijective. Hence there is an identification $G_p(V) = GL(V)/P_e$ such that ι becomes the identity and $\rho = \pi$. Observe that this identification is not intrinsic but depends on the base point e.

The isotropy group P_e can be computed explicitly. Pick $y \in G_{n-p-1}(V)$ such that e and y are independent. Then

$$GL(V) = \begin{pmatrix} GL(e) & L(y,e) \\ L(e,y) & GL(y) \end{pmatrix}$$

splits by

$$g = \begin{pmatrix} g_{11} & g_{12} \\ g_{21} & g_{22} \end{pmatrix}$$

$$= j_e \circ g_{11} \circ \pi_e + j_e \circ g_{12} \circ \pi_y + j_y \circ g_{21} \circ \pi_e + j_y \circ g_{22} \circ \pi_y \, .$$

Then $P_e = \{g \in GL(V) \mid g_{22} = 0\}$. The splitting depends on y.

The situation can be improved on a hermitian vector space. Let $\mathfrak{l}$ be a positive definite hermitian form on V. Define $(\mathfrak{x} \mid \mathfrak{y}) = \mathfrak{l}(\mathfrak{x}, \mathfrak{y})$ and $|\mathfrak{x}| = \sqrt{(\mathfrak{x} \mid \mathfrak{x})}$. A parabolic exhaustion $\tau: V \to \mathbb{R}_+$ is defined by $\tau(\mathfrak{x}) = |\mathfrak{x}|^2$. (See [31].) The form $\mathfrak{l}$ induces a Fubini-Study Kaehler metric on $\mathbb{P}(V)$ whose exterior form is denoted by ω. The exterior derivative $d = \partial + \bar{\partial}$ twists to $d^c = (i/4\pi)(\bar{\partial} - \partial)$. Then $\mathbb{P}^*(\omega) = dd^c \log \tau$ on $V - \{0\}$. If $x \in G_p(V)$, then

$$\int_{\ddot{E}(x)} \omega^p = 1 \qquad \int_{\mathbb{P}(V)} \omega^n = 1 \, .$$

The degree of a pure p-dimensional analytic subset A of $\mathbb{P}(V)$ is computed by

$$\int_A \omega^p = \deg A \, .$$

The hermitian metric $\mathfrak{l}$ induces hermitian metrics $\mathfrak{l}$ on V^*, $\underset{p}{\wedge} V$, $\underset{p}{\oplus} V$, $\underset{p}{\wedge} V$ and a Fubini-Study Kaehler metric on $\mathbb{P}(\underset{p+1}{\wedge} V)$ whose exterior form is denoted by ω_p. Then

$$D(p,n) = \deg G_p(V) = \int_{G_p(V)} (\omega_p)^{d(p,n)} .$$

For $x \in G_p(V)$, define $x^\perp \in G_{n-p-1}(V)$ by

$$E(x^\perp) = E(x)^\perp = \{ \mathfrak{z} \in V \mid (\mathfrak{z} \mid \mathfrak{x}) = 0 \ \forall\ \mathfrak{x} \in E(x) \} .$$

Then x and $x^\perp$ are independent. Abbreviate $L(x) = L(x,x^\perp)$.

A connected, compact, real Lie subgroup of $GL(V)$ is defined by

$$\mathfrak{U} = \mathfrak{U}(V) = \mathfrak{U}(V,\mathfrak{l}) = \{ g \in GL(V) \mid |g(\mathfrak{x})| = |\mathfrak{x}| \ \forall\ \mathfrak{x} \in GL(V) \} .$$

Here $\mathfrak{U}$ is called the *unitary group* of V for $\mathfrak{l}$. The group $\mathfrak{U}$ acts transitively on $G_p(V)$. For $x \in G_p(V)$, abbreviate $\mathfrak{U}(x) = \mathfrak{U}(E(x))$. Pick $e \in G_p(V)$. Then

$$GL(V)/P_e = \mathfrak{U}(V)/P_e \cap \mathfrak{U}(V)$$

are identified. The maps π and ρ restrict to $\pi : \mathfrak{U}(V) \to G_p(V)$ and $\rho : \mathfrak{U}(V) \to \mathfrak{U}(V)/P_e \cap \mathfrak{U}(V)$ with $\iota \circ \rho = \pi$. Pick $y = e^\perp$ for the splitting of $GL(V)$. Then

$$\mathfrak{U}(V) \cap P_e = \begin{pmatrix} \mathfrak{U}(e) & 0 \\ 0 & \mathfrak{U}(e^\perp) \end{pmatrix} = \mathfrak{U}(e) \times \mathfrak{U}(e^\perp)$$

$$\mathfrak{U}(V) \cap P_e = \{ g \in \mathfrak{U}(V) \mid g_{12} = 0 = g_{21} \} .$$

Again, take $y = e^\perp$. Then $k = j_e \circ \pi_e - j_y \circ \pi_y \in \mathfrak{U}(V)$ where $k \circ k$ is the identity on V. An inner automorphism σ of $\mathfrak{U}(V)$ is defined by $\sigma(g) = k \circ g \circ k$ for $g \in \mathfrak{U}(V)$. If $g \in \mathfrak{U}(V)$, then $k \circ g = g \circ k$ if and only if $g \in \mathfrak{U}(e) \times \mathfrak{U}(e^\perp)$. Hence $\mathfrak{U}(V) \cap P_e$ is the set of fixed points of σ.

Therefore $G_p(V)$ is a symmetric space. Observe that $g^*(\omega_p) = \omega_p$ for all $g \in \mathfrak{U}(V)$ where ω_p is the exterior form of the Fubini-Study Kaehler metric on $G_p(V)$.

A form χ of degree m is said to be invariant on $G_p(V)$ if $g^*(\chi) = \chi$ for all $g \in \mathfrak{U}(V)$. Since $G_p(V)$ is a symmetric space, χ is invariant if and only if χ is harmonic. Let $\mathrm{Inv}^m(G_p(V))$ be the set of invariant forms of degree m on $G_p(V)$. Then $\mathrm{Inv}^m(G_p(V)) = H^m(G_p(V), \mathbb{C})$ can be identified with the mth cohomology group under the de Rham isomorphism. In particular $d\chi = 0$ if χ is invariant, and $\chi = 0$ if $\chi = d\xi$ is invariant and total. (See Lascoux-Berger [20] V §2 and standard textbooks.)

b) *Flag spaces of symbol* $\mathfrak{a}$

Let p and n be integers with $0 \leq p \leq n$. An integral valued vector

$$\mathfrak{a} = (a_0, \cdots, a_p) \in \mathbf{Z}^{p+1}$$

is said to be a *symbol of type* (p,n) if $0 \leq a_0 \leq a_1 \leq \cdots \leq a_p \leq n-p$. Let $\mathfrak{S}(p,n)$ be the set of symbols of type (p,n). If $\mathfrak{a} \in \mathfrak{S}(p,n)$, define $\vec{\mathfrak{a}} = a_0 + \cdots + a_p$. Also define $\hat{a}_\mu = a_\mu + \mu$ for $\mu = 0, \cdots, p$ and $\hat{\mathfrak{a}} = (\hat{a}_0, \cdots, \hat{a}_p)$. Define

$$a_{-1} = 0 \qquad \hat{a}_{-1} = -1 \qquad a_{p+1} = n-p \qquad \hat{a}_{p+1} = n+1$$

$$\mathfrak{S}(p,n,m) = \{\mathfrak{a} \in \mathfrak{S}(p,n) \mid \vec{\mathfrak{a}} = m\} \qquad \text{if } 0 \leq m \in \mathbf{Z}.$$

If $\mathfrak{a} \in \mathfrak{S}(p,n)$, then $\mathfrak{a}^* = (n-p-a_p, \cdots, n-p-a_0) \in \mathfrak{S}(p,n)$. Also $\mathfrak{S}(p,n)$ is partially ordered by setting $\mathfrak{a} \leq \mathfrak{b}$ if and only if $a_\mu \leq b_\mu$ for $\mu = 0, \cdots, p$.

Let V be a complex vector space of dimension $n+1 > 0$. For $\mathfrak{a} \in \mathfrak{S}(p,n)$ define

$$G_{\mathfrak{a}}(V) = \prod_{q=0}^{p} G_{\hat{a}_q}(V).$$

The *flag space of symbol* $\mathfrak{A}$ is defined by

$$F(\mathfrak{A}) = \{(v_0,\cdots,v_p) \in G_{\mathfrak{A}}(V) \mid E(v_0) \subset \cdots \subset E(v_p)\} .$$

Then $F(\mathfrak{A})$ is a connected, compact, smooth, complex submanifold of $G_{\mathfrak{A}}(V)$ with dimension

$$d(\mathfrak{A}) = \dim F(\mathfrak{A}) = \sum_{q=0}^{p} (n-\hat{a}_q)(\hat{a}_q - \hat{a}_{q-1}) .$$

If $v \in G_{\mathfrak{A}}(V)$, then the coordinates of v are always denoted by $v_0,\cdots,v_p$ such that $v = (v_0,\cdots,v_p)$.

The general linear group $GL(V)$ acts on $G_{\mathfrak{A}}(V)$ with $g(F(\mathfrak{A})) = F(\mathfrak{A})$ for each $g \in GL(V)$. Then $GL(V)$ acts holomorphically and transitively on $F(\mathfrak{A})$ as a group of biholomorphic maps. If V is a hermitian vector space, $\mathfrak{U}(V)$ acts transitively on $F(\mathfrak{A})$.

Take $e \in F(\mathfrak{A})$. The isotropy group $P_e = \{g \in GL(V) \mid g(e) = e\}$ is a closed subgroup of $GL(V)$. Let $\rho(g) = gP_e$ be the left coset of $g \in GL(V)$ and let $GL(V)/P_e$ be the left coset space. A holomorphic fiber bundle $\pi : GL(V) \to F(\mathfrak{A})$ is defined by $\pi(g) = g(e)$. One and only one map $\iota : GL(V)/P_e \to F(\mathfrak{A})$ exists such that $\iota \circ \rho = \pi$. Here ι is bijective. Hence $F(\mathfrak{A}) = GL(V)/P_e$ can be identified such that ι becomes the identity and such that $\rho = \pi$. This identification is not intrinsic, but depends on the base point e.

The isotropy group P_e can be computed explicitly. Pick $c_\mu \in G_{\hat{a}_\mu - \hat{a}_{\mu-1}}(V)$ such that $e_\mu = c_0 \wedge \cdots \wedge c_\mu$ for $\mu = 0,\cdots,p+1$ with $e_{p+1} = \infty \in G_n(V)$. Then

$$V = E(c_0) \oplus E(c_1) \oplus \cdots \oplus E(c_{p+1}) .$$

Let $j_\mu : E(c_\mu) \to V$ be the inclusion and let $\pi_\mu : V \to E(c_\mu)$ be the projection for each $\mu = 0,\cdots,p+1$. For $g \in GL(V)$, define $g_{\mu\nu} = \pi_\mu \circ g \circ j_\nu \in L(c_\nu, c_\mu)$. Then

$$g = \sum_{\mu,\nu=0}^{p+1} j_\mu \circ g_{\mu\nu} \circ \pi_\nu \ .$$

Write $g = \text{matrix}(g_{\mu\nu})$. Then

$$P_e = \{g \in GL(V) \mid g_{\mu\nu} = 0 \ \forall \ \mu > \nu\} \ .$$

Let $\mathfrak{l}$ be a positive definite hermitian form on V . Then $\mathfrak{U} = \mathfrak{U}(V) = \mathfrak{U}(V,\mathfrak{l})$ is defined. Pick $e \in F(\mathfrak{a})$. Then

$$GL(V)/P_e = \mathfrak{U}(V)/P_e \cap \mathfrak{U}(V) \ .$$

Here $P_e \cap \mathfrak{U}(V)$ can be computed. Take $c_0 = e_0$ and $e_\mu = e_{\mu-1}^\perp$ in $E(e_\mu)$ for $\mu = 1,\cdots,p+1$. Then $e_\mu = c_0 \wedge \cdots \wedge c_\mu$. The direct sum decomposition of V is orthogonal. Therefore

$$\mathfrak{U}(V) \cap P_e = \{g \in \mathfrak{U}(V) \mid g_{\mu\nu} = 0 \ \text{ if } \ \mu \neq \nu\}$$

$$\mathfrak{U}(V) \cap P_e = \mathfrak{U}(c_0) \times \cdots \times \mathfrak{U}(c_{p+1})$$

and $\mathfrak{U}(V) \cap P_e$ is a compact, connected subgroup of $\mathfrak{U}(V)$.

For this section compare Chern [5], Chern [7], Cowen [8], Matsushima [21] and Vesentini [34].

c) *Short flags*

Let p and q be integers with $0 \leq q \leq p \leq n$. Define

$$F_{pq} = \{(x,y) \in G_p(V) \times G_q(V) \mid E(y) \subseteq E(x)\} \ .$$

If $p = q$, then F_{pp} is the diagonal of $G_p(V) \times G_p(V)$. If $p > q$, then F_{pq} is biholomorphically equivalent to $F(q,p-1)$ under the map $(x,y) \to (y,x)$. Hence F_{pq} is a connected, compact, smooth, complex submanifold of $G_p(V) \times G_q(V)$ with

$$\dim F_{pq} = d(p,n) + (p-q)(q+1) = d(q,n) + (n-p)(p-q) \ .$$

Let $\pi : F_{pq} \to G_q(V)$ and $\tau : F_{pq} \to G_p(V)$ be the projections. Since $\pi^{-1}(y)$ is isomorphic to $G_{p-q-1}(V/E(y))$, the fiber $\pi^{-1}(y)$ is a connected, compact, complex manifold of dimension $(n-p)(p-q)$ for each $y \in G_q(V)$. Since $\tau^{-1}(x)$ is isomorphic to $G_q(E(x))$, the fiber $\tau^{-1}(x)$ is a connected, compact, complex manifold of dimension $(p-q)(q+1)$. If $g \in GL(V)$, then $g(F_{pq}) = F_{pq}$ and $GL(V)$ acts holomorphically and transitatively on F_{pq} with $g \circ \pi = \pi \circ g$ and $g \circ \tau = \tau \circ g$. Hence π and τ are regular and surjective holomorphic maps. It is easy to show directly that π and τ are locally trivial. This fact is also a consequence of the following lemma which is needed for other purposes.

LEMMA 1.1. *Let* K *be a closed complex Lie subgroup of the complex Lie group* G. *Let* $M = G/K$ *be the left coset space. Define* $\rho : G \to M$ *by* $\rho(g) = gK$. *Then* M *is a complex manifold and* $\rho : G \to M$ *is a holomorphic fiber bundle. Let* N *be a complex space and let* $\pi : N \to M$ *be a holomorphic map. Assume that* G *acts holomorphically on* N *as a group of biholomorphic maps such that* $\pi \circ g = g \circ \pi$ *for all* $g \in G$. *Then* π *is locally trivial and surjective.*

Proof. The group G acts transitively on M by $h \cdot \rho(g) = \rho(hg)$. Define $e = \rho(K)$. As known, M is a complex manifold and ρ is a holomorphic fiber bundle. Moreover, an open neighborhood U of e and a holomorphic section $s : U \to G$ exist such that $\rho \circ s$ is the identity on U (Hirzebruch [16] Satz 3.4.3). Then $s(x)e = x$ if $x \in U$. Define $\tilde{U} = \pi^{-1}(U)$ and $F = \pi^{-1}(e)$. If $z \in \tilde{U}$, then $s(\pi(z))(e) = \pi(z)$. Hence $e = s(\pi(z))^{-1}\pi(z)$. Since π commutes with the action of G, a holomorphic map $\beta : \tilde{U} \to U \times F$ is defined by $\beta(z) = (\pi(z), s(\pi(z))^{-1}z)$ for all $z \in \tilde{U}$. Let $\psi : U \times F \to U$ be the projection. If $(x,y) \in U \times F$, then $\pi(s(x)y) = s(x)\pi(y) = s(x)e \in U$. Hence $s(x)y \in \tilde{U}$ and a holomorphic map $\alpha : U \times F \to \tilde{U}$ is defined by $\alpha(x,y) = s(x)y$. Observe $\pi \circ \alpha = \psi$ and $\psi \circ \beta = \pi$. Then $\beta \circ \alpha : U \times F \to U \times F$ and $\alpha \circ \beta : \tilde{U} \to \tilde{U}$ are identities. Hence $\alpha : U \times F \to \tilde{U}$ is biholomorphic and provides a local trivialization of π over U.

Take $x_0 \in M$. Then there exists $g \in G$ such that $g(e) = x_0$. Then $U_g = g(U)$ and $\tilde{U}_g = g(\tilde{U})$ are open in M and N respectively with $x_0 \in U_g$ and $\tilde{U}_g = \pi^{-1}(U_g)$. Define $F_g = g(F)$. Then $F_g = \pi^{-1}(x_0)$. A biholomorphic map $\tilde{g}: U \times F \to U \times F_g$ is defined by $\tilde{g}(x,y) = (g(x), g(y))$. Also $g: \tilde{U} \to \tilde{U}_g$ is biholomorphic. The map

$$\alpha_g = g \circ \alpha \circ \tilde{g}^{-1} : U_g \times F_g \to \tilde{U}_g$$

is biholomorphic and a trivialization of π over U_g. Hence π is surjective and locally trivial; q.e.d.

If $q = 0$, write $F_p = F_{p,0}$. Then

$$F_p = \{(x,y) \in G_p(V) \times P(V) \mid y \in \ddot{E}(x)\} . \tag{1.1}$$

Then F_p is an admissible family in the sense of [27]. The flag space F_p can also be obtained by another method: The *tautological bundle*

$$S_p(V) = \{(x, \mathfrak{z}) \in G_p(V) \times V \mid \mathfrak{z} \in E(x)\} \tag{1.2}$$

is a holomorphic sub-bundle of the trivial vector bundle $G_p(V) \times V$, with the quotient bundle $Q_p(V)$. A short exact sequence

$$0 \longrightarrow S_p(V) \xrightarrow[j]{} G_p(V) \times V \xrightarrow[\eta]{} Q_p(V) \longrightarrow 0 \tag{1.3}$$

is defined where j is the inclusion. This defines projective bundles $P(S_p(V))$ and $P(Q_p(V))$ and an inclusion

$$j : P(S_p(V)) \to G_p(V) \times P(V) .$$

In fact $P(S_p(V)) = F_p$ by (1.1).

d) *Analyticity criteria*

Subsequently, certain sets will be defined geometrically. The following criteria will be helpful in establishing their analytic properties.

LEMMA 1.2. *Let* M *and* N *be complex spaces. Assume that* N *is irreducible and has dimension* n. *Let* $f: M \to N$ *be a proper, surjective, q-fibering holomorphic map such that* $f^{-1}(y)$ *is irreducible for all* $y \in N$. *Then* M *is irreducible with* $\dim M = n+q$.

Proof. Let S be the set of singular points of M. Take any branch B of M. The restriction $g = f|B: B \to N$ is proper.

CLAIM 1. If $a \in B-S$ and $b = f(a)$, then $f^{-1}(b) = g^{-1}(b)$.

Proof of Claim 1. Obviously $g^{-1}(b) = f^{-1}(b) \cap B \subseteq f^{-1}(b)$, and $a \in f^{-1}(b) \cap (B-S) = g^{-1}(b) \cap (B-S)$. Since $B-S$ is open in M, the fiber $g^{-1}(b)$ has dimension q at a. Since $f^{-1}(b)$ is irreducible and q-dimensional, $f^{-1}(b) = g^{-1}(b)$. Claim 1 is proved.

CLAIM 2. The map $f: B-S \to N$ is open.

Proof of Claim 2. Let U be open in $B-S$. Assume that $f(U)$ is not open. Then a point $b \in f(U)$ and a sequence $\{y_p\}_{p \in N}$ of points $y_p \in N - f(U)$ exist such that $y_p \to b$ for $p \to \infty$. Here $y_p \neq y_q$ if $p \neq q$ can be assumed. Observe that $y_p \neq b$ for all $p \in N$. The set $Y = \{y_p \mid p \in N\}$ is analytic in $N - \{b\}$. Hence $F = f^{-1}(Y)$ is analytic in $M - f^{-1}(b)$. Then $\{f^{-1}(y_p)\}_{p \in N}$ is the set of branches of F. Hence F is pure q-dimensional. Take $a \in f^{-1}(b) \cap U$. Now $f^{-1}(b)$ is irreducible and pure q-dimensional, and $a \notin \overline{F}$. Thus, by the Remmert-Stein Theorem, the set $\overline{F}$ is analytic in M. Moreover the closures of the branches of F are exactly the branches of $\overline{F}$. Since each $f^{-1}(y_p)$ is compact and contained in $M - f^{-1}(b)$, the family $\{f^{-1}(y_p)\}_{p \in N}$ is the family of branches of $\overline{F}$. Hence $F = \overline{F}$. By the proper mapping theorem, $Y = f(F) = f(\overline{F})$ is analytic in N which is false. Hence $f(U)$ is open. Claim 2 is proved.

CLAIM 3. The map $g: B \to N$ is surjective and $\dim B = n+q$.

Proof of Claim 3. g is proper. Hence $g(B)$ is analytic and $g(B-S) \neq \emptyset$ is open in N. Since N is irreducible, $g(B) = N$. For the definition and the properties of the rank of a holomorphic map see [1] §1. The set $E = \{x \in B \mid \mathrm{rank}_x g < n\}$ is analytic and $g(E)$ is thin in N. Hence $E \neq B$. Take $a \in B - (E \cup S)$. Define $b = f(a)$. Then

$$n = \mathrm{rank}_a g = \dim B - \dim_a g^{-1}(b) = \dim B - q .$$

Therefore $\dim B = n+q$. Claim 3 is proved.

Take $z \in M$. Define $y = f(z)$. Take $x \in g^{-1}(y) \subseteq B$. Then

$$n \geq \mathrm{rank}_x g = n+q - \dim_x g^{-1}(y) .$$

Hence $\dim_x g^{-1}(y) \geq q$. Because $f^{-1}(y)$ is irreducible, q-dimensional and contains $g^{-1}(y)$, we have $g^{-1}(y) = f^{-1}(y)$. Hence $z \in g^{-1}(y) \subseteq B$. Therefore $M = B$ is irreducible and $(n+q)$-dimensional; q.e.d.

LEMMA 1.3. *Let* M *be an irreducible complex space of dimension* m. *Let* N *be a connected, simply connected, complex manifold of dimension* n *with* $m-n = q \geq 0$. *Let* $f: M \to N$ *be a surjective, locally trivial, holomorphic map. Then* $f^{-1}(y)$ *is irreducible and* q*-dimensional for all* $y \in N$.

Proof. Because f is locally trivial, f is open and q-fibering. At first, *assume* $f^{-1}(y)$ *has disjoint branches for each* $y \in Y$. For each $y \in Y$, let $L(y) \neq \emptyset$ be the set of branches of $f^{-1}(y)$. Let

$$L = \bigcup_{y \in N} \{y\} \times L(y)$$

be the disjoint union. Define $h: L \to N$ by $h(y,z) = y$. For each $x \in M$, let $g_0(x) \in L(f(x))$ be the branch of $f^{-1}(f(x))$ containing x. The map $g: M \to L$ defined by $g(x) = (f(x), g_0(x))$ is surjective with $h \circ g = f$.

Introduce the quotient topology on L. Then L is arcwise connected and g is continuous. If U is open in N, then $f^{-1}(U) = g^{-1}(h^{-1}(U))$ is open. Therefore h is continuous. Let V be open in L.

Then $U = g^{-1}(V)$ is open in M with $g(U) = V$. Because f is open, the set $h(V) = h(g(U)) = f(U)$ is open in N. Hence h is open. Trivially, h is surjective.

Take $b \in N$. Then $L(b) \neq \emptyset$ is at most countable. Therefore $p \leq \infty$ exists such that $L(b) = \{F_\mu \mid \nu \in \mathbf{Z}[1,p]\}$ with $F_\nu \neq F_\mu$ for $\mu \neq \nu$. Then $F = f^{-1}(b)$ is the disjoint union of the F_ν with $\nu \in \mathbf{Z}[1,p]$. Because f is locally trivial, there exists an open, connected neighborhood U of b and a biholomorphic map $\alpha : V \to F \times U$ such that $\pi \circ \alpha = f$, where $V = f^{-1}(U)$ and $\pi : F \times U \to U$ is the projection. Also we can assume $\alpha(x) = (x,b)$ for all $x \in F$. Then $V_\nu = \alpha^{-1}(F_\nu \times U)$ is open and connected in V, and V is the disjoint union of the V_ν for $\nu \in \mathbf{Z}[1,p]$. Take $y \in U$. Then

$$V_\nu \cap f^{-1}(y) = F_\nu(y) = \alpha^{-1}(F_\nu \times \{y\})$$

is a branch of $f^{-1}(y)$ with $F_\nu(y) \neq F_\mu(y)$ for $\nu \neq \mu$. Hence

$$L(y) = \{F_\nu(y) \mid \nu \in \mathbf{Z}[1,p]\} .$$

The set $W_\nu = g(V_\nu) = \{(y, F_\nu(y)) \mid y \in U\}$ is connected with $g^{-1}(W_\nu) = V_\nu$. Hence W_ν is open with $(b,F_\nu) \in W_\nu$. If $\nu \neq \mu$, then $W_\nu \cap W_\mu = \emptyset$. The union

$$W = \bigcup_{\nu=1}^{p} W_\nu$$

is disjoint and $V = g^{-1}(W)$. The map $h : W_\nu \to U$ is bijective, open, and continuous, and is therefore a homeomorphism. Observe that $h^{-1}(U) = W$. Therefore, provided that L is a Hausdorff space, $h : L \to W$ is a covering space.

Take c_1 and c_2 in L with $c_1 \neq c_2$. If $h(c_1) \neq h(c_2)$, open neighborhoods A_μ of $h(c_\mu)$ exist such that $A_1 \cap A_2 = \emptyset$. Then $B_\mu = h^{-1}(A_\mu)$ is an open neighborhood of c_μ in L with $B_1 \cap B_2 = \emptyset$. If $h(c_1) = h(c_2) = b$, the previous construction holds. One and only one W_{ν_μ} with $c_\mu \in W_{\nu_\mu}$ exists for $\mu = 1,2$. Because $h|W_{\nu_\mu}$ is injective,

$\nu_1 \neq \nu_2$. Therefore $W_{\nu_1} \cap W_{\nu_2} = \emptyset$. Hence L is a Hausdorff space and $h: L \to N$ is a covering space. Since L and N are connected and since L is simply connected, $h: L \to N$ is a homeomorphism. Therefore $\#L(y) = 1$ for all $y \in N$. The lemma is proved in this special case.

Consider the general case. Let S be the set of singular points of M. Define $M_0 = M - S$ and $f_0 = f|M_0$. For each $y \in N$ let $L(y)$ (respectively $L_0(y)$) be the set of branches of $f^{-1}(y)$ (respectively $f_0^{-1}(y)$). Take $b \in M$. Define $F = f^{-1}(b)$. Then $L(b) = \{F_\nu \mid \nu \in Z[1,p]\}$ with $p \leq \infty$ and with $F_\nu \neq F_\mu$ if $\nu \neq \mu$. There exists an open, connected neighborhood U of b and a biholomorphic map $\alpha: V \to F \times U$ such that $\pi \circ \alpha = f$, where $V = f^{-1}(U)$ and $\pi: F \times U \to U$ is the projection. Moreover we can assume that $\alpha(x) = (x,b)$ for all $x \in F$. Let T be the singular set of F. Then $T \times U$ is the singular set of $F \times U$. Hence $V \cap S = \alpha^{-1}(T \times U)$ and $F \cap S = T$. Define $F_\nu^0 = F_\nu - T = F_\nu - S = F_\nu \cap M_0 \neq \emptyset$. Then $L^0(b) = \{F_\nu^0 | \nu \in Z[1,p]\}$ where $F_\nu^0 \neq F_\mu^0$ for $\nu \neq \mu$. Therefore the following three consequences have been established:

1) If $y \in N$, then $\#L(y) = \#L_0(y)$.
2) The holomorphic map $f_0: M_0 \to N$ is surjective and locally trivial.
3) For each $y \in N$, the branches of $f_0^{-1}(y)$ are disjoint.

Consequently, $\#L_0(y) = 1$ for all $y \in N$. Hence $\#L(y) = 1$ for all $y \in N$. Therefore $f^{-1}(y)$ is irreducible and q-dimensional for each $y \in N$; q.e.d.

These Lemmata will be quite useful. The first already helps in proving the following result:

LEMMA 1.4. *Take* $p_\mu \in Z[0,n]$ *for* $\mu = 1, \cdots, k$. *Then*

$$N_k = \left\{ (x, v_1, \cdots, v_k) \in P(V) \times \prod_{\mu=1}^{k} G_{p_\mu}(V) \,\middle|\, x \in \bigcap_{\mu=1}^{k} \ddot{E}(v_\mu) \right\}$$

is an irreducible analytic subset of $Y_k = P(V) \times \prod_{\mu=1}^{k} G_{p_\mu}(V)$.

Proof. If $k=1$, then N_1 is biholomorphically equivalent to F_{p_1} under the map $(x,y) \to (y,x)$. Hence N_1 is an analytic subset of $P(V) \times G_{p_1}(V)$. As a flag space, F_{p_1} is a connected manifold. Hence N_1 is irreducible. (Lemma 1.2 also could have been applied, because $\tau : F_p \to G_p(V)$ is a surjective, p-fibering map with irreducible fibers $\ddot{E}(x)$.)

Now, under the induction hypothesis for $k-1$, the lemma will be proved for k. Then N_{k-1} is irreducible. Let $\rho : N_{k-1} \to P(V)$ and $\pi : F_{p_k} \to P(V)$ be the projection. Let $(\tilde{N}_k, \tilde{\rho}, \tilde{\pi})$ be the relative product of (ρ, π) (see Tung [33] §8.1). Take the standard model. Then

$$\tilde{N}_k = \{(z,w) \in N_{k-1} \times F_{p_k} \mid \rho(z) = \pi(w)\}$$
$$= \left\{((x,v),(v_k,x)) \in Y_{k-1} \times (G_{p_k}(V) \times P(V) \,\middle|\, x \in \bigcap_{\mu=1}^{k} \ddot{E}(v_\mu)\right\},$$

where $v = (v_1, \cdots, v_{k-1})$. Here $\tilde{N}_k$ is an analytic subset. Define

$$\Delta = \{((x_1 v_1, \cdots, v_{k-1}), (v_k, x)) | v_\mu \in G_{p_\mu}(V) \text{ and } x \in P(V)\}.$$

Then Δ is analytic in $Y_{k-1} \times G_k(V) \times P(V)$ with $\tilde{N}_k \subseteq \Delta$. A biholomorphic map $j : \Delta \to Y_k$ is defined by

$$j((x_1 v_1, \cdots, v_{k-1}), (v_k, x)) = (x_1 v_1, \cdots, v_k).$$

Then $N_k = j(\tilde{N}_k)$ is analytic in Y_k. Let $\hat{\pi} : N_k \to N_{k-1}$ and $\hat{\rho} : N_k \to F_{p_k}$ be the projections. Therefore the diagram (1.4) commutes.

(1.4)

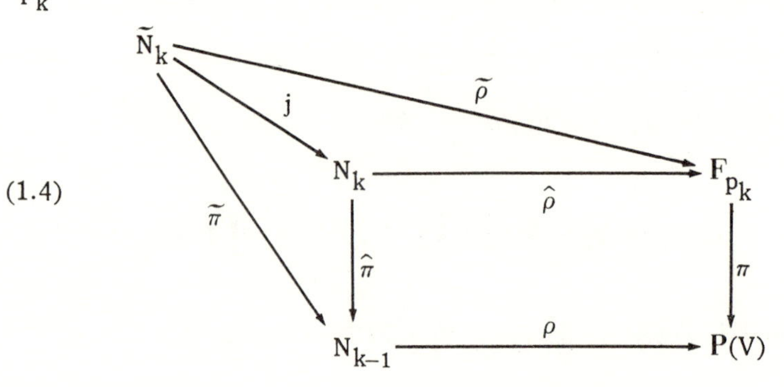

Hence $(N_k, \hat{\rho}, \hat{\pi})$ is another model of the relative product of (ρ, π). In particular, $\hat{\pi}^{-1}(y)$ is biholomorphically equivalent to $\pi^{-1}(\rho(y))$. Therefore the fibers of $\hat{\pi}$ are irreducible and have pure dimension $(n-p)p$. Also $\hat{\pi}$ is locally trivial and surjective. By the induction assumption N_{k-1} is irreducible. Hence N_k is irreducible; q.e.d.

LEMMA 1.5. *Take* $p_\mu \in \mathbf{Z}[0,n]$ *for* $\mu = 1, \cdots, k$. *Take* $0 \le q \in \mathbf{Z}$. *Define* $X_k = G_{p_1}(V) \times \cdots \times G_{p_k}(V)$. *Then*

$$M_q = \left\{ (v_1, \cdots, v_k) \in X_k \;\middle|\; \dim \bigcap_{\mu=1}^{k} E(v_\mu) \ge q \right\}$$

is an analytic subset of X_k.

Proof. Define N_k as in Lemma 1.4. The projection $f: N_k \to X_k$ is proper and holomorphic. The lemma is trivial if $q = 0$. Assume $q > 0$. By Remmert, the set

$$\begin{aligned} L_q &= \{ z \in N_k \mid \dim_z f^{-1}(f(z)) \ge q-1 \} \\ &= \{ z \in N_k \mid \operatorname{rank}_z f \le \dim N_k - q + 1 \} \end{aligned}$$

is analytic (see [1] §1). Therefore $f(L_q)$ is analytic. Take $v = (v_1, \cdots, v_k) \in X_k$. Then $v \in f(L_q)$ if and only if $x \in \mathbf{P}(V)$ exists such that $(x,v) \in L_q$ with $\dim_{(x,v)} f^{-1}(v) \ge q-1$, which is the case if and only if

$$\dim_x \bigcap_{\mu=1}^{k} \ddot{E}(v_\mu) \ge q-1 . \tag{1.5}$$

Since $\ddot{E}(v_1) \cap \cdots \cap \ddot{E}(v_k) = \ddot{E}$ is a projective plane, $\ddot{E}$ is irreducible. Hence $v \in f(L_q)$ if and only if $\dim \ddot{E} \ge q-1$, which holds if and only if

$$\dim \bigcap_{\mu=1}^{k} E(v_\mu) \ge q .$$

Therefore $M_q = f(L_q)$ is analytic in X_k; q.e.d.

A subset U of a complex space M is said to be *Zariski open* in M if $M-U$ is analytic. A Zariski open subset is open. If M is irreducible and if $U \neq \emptyset$ is Zariski open in M, then U is dense in M.

2. SCHUBERT VARIETIES

a) *Schubert families*

Let V be a complex vector space of dimension $n+1>0$. Take $p \in \mathbf{Z}[0,n]$ and $\mathfrak{a} \in \mathfrak{S}(p,n)$. The *Schubert family* $S(\mathfrak{a})$ of symbol a is the set of all $(x,v) \in G_p(V) \times F(\mathfrak{a})$ with $v = (v_0,\cdots,v_p)$ such that

$$\text{(2.1)} \qquad \dim E(x) \cap E(v_\mu) \geq \mu+1 \qquad \text{for } \mu = 0,1,\cdots,p \ .$$

Now Lemma 1.5 implies easily that $S(\mathfrak{a})$ is an analytic subset of $G_p(V) \times F(\mathfrak{a})$. Define $b_\mu = \hat{a}_\mu = a_\mu + \mu$ for $\mu = 0,\cdots,p$. Let $(\mathfrak{n}_0,\cdots,\mathfrak{n}_n)$ be a base of V. Define

$$x = P(\mathfrak{n}_{b_0} \wedge \cdots \wedge \mathfrak{n}_{b_p}) \qquad e_\mu = P(\mathfrak{n}_0 \wedge \cdots \wedge \mathfrak{n}_{b_\mu})$$

for $\mu = 0,\cdots,p$. Then $(x, e_0,\cdots,e_p) \in S(\mathfrak{a})$. Hence $S(\mathfrak{a}) \neq \emptyset$.

The general linear group $GL(V)$ acts holomorphically on $G_p(V) \times F(\mathfrak{a})$ with $g(S(\mathfrak{a})) = S(\mathfrak{a})$ for all $g \in GL(V)$. Hence $GL(V)$ acts holomorphically on $S(\mathfrak{a})$ as a group of biholomorphic maps. The action is not transitive in general. Let

$$\text{(2.2)} \qquad \pi : S(\mathfrak{a}) \to G_p(V) \qquad \sigma : S(\mathfrak{a}) \to F(\mathfrak{a})$$

be the projections. Then $g \circ \pi = \pi \circ g$ and $g \circ \sigma = \sigma \circ g$ for all $g \in GL(V)$. The representations of $G_p(V)$ and $F(a)$ in §1 and Lemma 1.1 imply: (compare Cowen [8]).

LEMMA 2.1. *The holomorphic maps* $\pi : S(\mathfrak{a}) \to G_p(V)$ *and* $\sigma : S(\mathfrak{a}) \to F(\mathfrak{a})$ *are locally trivial and surjective.*

For each $x \in G_p(V)$ and $v \in F(\mathfrak{a})$, the sets

$$(2.3) \qquad S(v,\mathfrak{a}) = \pi(\sigma^{-1}(x)) \qquad S_x(\mathfrak{a}) = \sigma(\pi^{-1}(x))$$

are analytic. The restrictions $\pi : \sigma^{-1}(v) \to S(v,\mathfrak{a})$ and $\sigma : \pi^{-1}(x) \to S_x(\mathfrak{a})$ are biholomorphic. By Lemma 2.1, $\{S(v,\mathfrak{a})\}_{v \in F(\mathfrak{a})}$ and $\{S_x(\mathfrak{a})\}_{x \in G_p(V)}$ are admissible families in the sense of Tung [33]. The analytic subset $S(v,\mathfrak{a})$ is called the *Schubert variety* for the flag v and the symbol $\mathfrak{a}$. Now, some examples shall be considered (compare Chern [7] §8).

EXAMPLE 1. $S(v,n-p,\cdots,n-p) = G_p(V)$.

Proof. Take $x \in G_p(V)$. Then $\dim E(v_\mu) \geq n-p+\mu+1$ and $\dim E(x) \geq p+1$ imply that $\dim E(x) \cap E(v_\mu) \geq \mu+1$; q.e.d.

EXAMPLE 2. $S(v,0,\cdots,0) = \{v_p\}$.

Proof. Observe $\dim E(v_\mu) = \mu+1$ for $\mu = 0,1,\cdots,p$. If $x \in G_p(V)$, then $\dim E(x) \cap E(v_\mu) \geq \mu+1$ for $\mu = 0,1,\cdots,p$ if and only if $x = v_p$; q.e.d.

REMARK. If $x \in S(v,\mathfrak{a})$, then $E(x) \subseteq E(v_p)$.

Proof. $\dim E(x) = p+1$ and $\dim E(x) \cap E(v_p) \geq p+1$ imply that $E(x) \subseteq E(v_p)$, q.e.d.

EXAMPLE 3. Take $r \in Z[-1,p]$. Define $a_\mu = 0$ if $0 \leq \mu \leq r$ and $a_\mu = 1$ if $r < \mu \leq p$. Define $E(v_{-1}) = 0$. Then

$$S(v,\mathfrak{a}) = \{x \in G_p(V) \mid E(v_r) \subseteq E(x) \subseteq E(v_p)\}.$$

Proof. Take $x \in S(v,\mathfrak{a})$. Then $E(x) \subseteq E(v_p)$. Also $\dim E(v_r) = r+1$ and $\dim E(x) \cap E(v_r) \geq r+1$ imply that $E(x) \supseteq E(v_r)$. Take $x \in G_p(V)$ with $E(v_r) \subseteq E(x) \subseteq E(v_p)$. If $0 \leq \mu \leq r$, then $E(v_\mu) \subseteq E(x)$. Hence $\dim E(x) \cap E(v_\mu) = \dim E(v_\mu) = \mu+1$. If $r < \mu \leq p$, then

$$\begin{aligned} \dim E(x) \cap E(v_\mu) &\geq \dim E(x) + \dim E(v_\mu) - \dim E(v_p) \\ &= (p+1) + (\mu+1) - (0+2) = \mu+1. \end{aligned}$$

Hence $x \in S(v,\mathfrak{a})$; q.e.d.

EXAMPLE 4. $S(v,1,\cdots,1) = \{x \in G_p(V) \mid E(x) \subseteq E(v_p)\}$.

Consider the short flag manifold $F_{p+1,p}$ with projections $\tau : F_{p+1,p} \to G_{p+1}(V)$ and $\pi : F_{p+1,p} \to G_p(V)$. Then $S(v,1,\cdots,1) = \pi(\tau^{-1}(v_p))$.

EXAMPLE 5. If $0 \le q < p$, then

$$S(v,a_0,\cdots,a_q,n-p,\cdots,n-p) = \{x \in G_p(V) \mid \dim E(x) \cap E(v_\mu) \ge \mu+1 \ \forall \ \mu = 0,\cdots,q\} .$$

Proof. Take $x \in G_p(V)$ with $\dim E(x) \cap E(v_\mu) \ge \mu+1$ for all $\mu \in \mathbf{Z}[0,q]$. Take $\mu \in \mathbf{Z}[q+1,p]$. Then

$$\dim E(x) \cap E(v_\mu) \ge p+1+n-p+\mu+1-(n+1) = \mu+1 .$$

Hence $x \in S(v,a_0,\cdots,a_q,n-p,\cdots,n-p)$; q.e.d.

REMARK. In the case of Example 5, define $\mathfrak{b} = (a_0,\cdots,a_q) \in \mathfrak{S}(q,n)$. Let $\phi : F(\mathfrak{a}) \to F(\mathfrak{b})$ be the projection $\phi(v_0,\cdots,v_p) = (v_0,\cdots,v_q)$. Define $\sigma_0 = \phi \circ \sigma : S(\mathfrak{a}) \to F(\mathfrak{b})$. Then σ_0 is surjective with $g \circ \sigma_0 = \sigma_0 \circ g$ for all $g \in GL(V)$. By Lemma 1.1 σ_0 is locally trivial.

EXAMPLE 6. $S(v,a_0,n-p,\cdots,n-p) = \{x \in G_p(V) \mid E(x) \cap E(v_0) \neq \emptyset\}$.

Consider the short flag manifold $G_p(V) \xleftarrow{\tau} F_p \xrightarrow{\pi} \mathbf{P}(V)$. Then $\ddot{E}(v_0) \subseteq \mathbf{P}(V)$ and

$$\tau\pi^{-1}(v_0) = S(v,a_0,n-p,\cdots,n-p) . \tag{2.4}$$

b) *Schubert cells*

Let $F^n = F(0,\cdots,0)$ be the complete flag space ($n+1$ zeroes). Take $\mathfrak{a} \in \mathfrak{S}(p,n)$. Let $S\langle\mathfrak{a}\rangle$ be the set of all $(x,v) \in G_p(V) \times F^n$ such that $\dim E(x) \cap E(v_{\hat{a}_\mu}) \ge \mu+1$ for $\mu = 0,1,\cdots,p$. Lemma 1.5 implies easily that $S\langle\mathfrak{a}\rangle$ is an analytic subset of $G_p(V) \times F^n$. Consider the commutative diagram (2.5), where all maps are the natural projections.

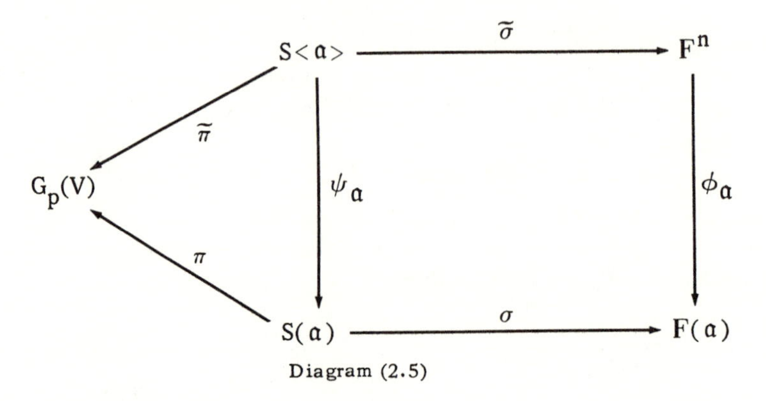

Diagram (2.5)

For $v \in F^n$ define

$$S\langle v, \alpha\rangle = \tilde{\pi}(\tilde{\sigma}^{-1}(v)) = S(\phi_\alpha(v), \alpha)\,. \tag{2.6}$$

Take $\alpha = (a_0, \cdots, a_p) \in \mathfrak{S}(p,n)$. Take $j \in \mathbf{Z}[0,p]$. Recall that $a_{-1} = 0$. If $a_{j-1} < a_j$, define

$$\partial_j \alpha = (a_0, \cdots, a_{j-1}, a_j - 1, a_{j+1}, \cdots, a_p) \in \mathfrak{S}(p,n) \tag{2.7}$$

and $\partial_j S\langle\alpha\rangle = S\langle\partial_j \alpha\rangle$. If $a_{j-1} = a_j$, define $\partial_j S\langle\alpha\rangle = \emptyset$. If $v \in F^n$, define $\partial_j S\langle v, \alpha\rangle = S\langle v, \partial_j \alpha\rangle$ if $a_{j-1} < a_j$, and $\partial_j S\langle v, \alpha\rangle = \emptyset$ if $a_{j-1} = a_j$. Observe

$$\partial_j S\langle v, \alpha\rangle \subseteq S\langle v, \alpha\rangle \qquad \partial_j S\langle\alpha\rangle \subseteq S\langle\alpha\rangle\,. \tag{2.8}$$

Define

$$S^*\langle v, \alpha\rangle = S\langle v, \alpha\rangle - \bigcup_{j=0}^{p} \partial_j S\langle v, \alpha\rangle \tag{2.9}$$

$$S^*\langle\alpha\rangle = S\langle\alpha\rangle - \bigcup_{j=0}^{p} \partial_j S\langle\alpha\rangle\,. \tag{2.10}$$

Here $S^*\langle v, \alpha\rangle$ is called the *open Schubert cell* for the flag v and the symbol a. Observe that $S^*\langle v, \alpha\rangle$ and $S^*\langle\alpha\rangle$ are Zariski open in $S\langle v, \alpha\rangle$ respectively $S\langle\alpha\rangle$. *The reader should note that* $\partial_j S(\alpha)$,

$S^*(\mathfrak{a})$, $\partial_j S(v,\mathfrak{a})$ and $S^*(v,\mathfrak{a})$ *do not make sense.* If $v \in F^n$ and $w = \phi_{\mathfrak{a}}(v)$, then $S\langle v,\mathfrak{a}\rangle = S(w,\mathfrak{a})$ and $S^*\langle v,\mathfrak{a}\rangle$ is Zariski open in $S(w,\mathfrak{a})$, but $S^*\langle v,\mathfrak{a}\rangle$ is *not intrinsically defined* by w and $\mathfrak{a}$ alone.

LEMMA 2.2. *Take* $v \in F^n$ *and* $\mathfrak{a} \in \mathfrak{S}(p,n)$. *Then*

$$S^*\langle v,\mathfrak{a}\rangle = \bigcap_{\mu=0}^{p} \{x \in S\langle v,\mathfrak{a}\rangle \mid \dim E(x) \cap E(v_{\hat{a}_\mu - 1}) = \mu\}.$$

If $x \in S^*\langle v,\mathfrak{a}\rangle$, *then* $\dim E(x) \cap E(v_{\hat{a}_\mu}) = \mu + 1$.

Proof. Abbreviate $b_\mu = \hat{a}_\mu = a_\mu + \mu$ for $\mu = 0,\cdots,p$. Take $x \in S^*\langle v,\mathfrak{a}\rangle$. Define $m_\mu = \dim E(x) \cap E(v_{b_\mu})$ and $n_\mu = \dim E(x) \cap E(v_{b_\mu - 1})$. Then $m_\mu \geq \mu+1$ and $0 \leq m_\mu - n_\mu \leq 1$. Now $m_\mu = \mu+1$ and $n_\mu = \mu$ have to be proved. The proof proceeds by induction for μ.

Assume $\mu = 0$. At first consider the case $a_0 = 0$. Then $b_0 = 0$ and $1 = \dim E(v_{b_0}) \geq m_0 \geq 1$. Therefore $m_0 = 1$. Also $b_0 - 1 = -1$, hence $n_0 = 0$. Now, consider the case $a_0 = b_0 > 0$. Assume $n_0 > 0$. This and $m_\mu \geq \mu+1$ for $\mu \geq 1$ imply $x \in S\langle v, \partial_0 \mathfrak{a}\rangle$ which is wrong. Hence $n_0 = 0$, which implies $m_0 \leq 1$. Therefore $m_0 = 1$. The case $\mu = 0$ is proved.

Assume $\mu \geq 1$ and assume that the cases $0,1,\cdots,\mu-1$ are proved. At first consider the case $a_{\mu-1} = a_\mu$. Then $b_\mu - 1 = b_{\mu-1}$. Hence $n_\mu = m_{\mu-1} = \mu$. Also $m_\mu \leq n_\mu + 1 = \mu+1$. Therefore $m_\mu = \mu+1$. Now consider the case $a_{\mu-1} < a_\mu$. Assume $n_\mu > \mu$. Then $n_\mu \geq \mu+1$. Also $m_\nu \geq \nu+1$ if $\nu \neq \mu$ and $\nu \in Z[0,p]$. Hence $x \in S\langle v, \partial_\mu \mathfrak{a}\rangle$ which is wrong. Hence $n_\mu = \mu$ and $m_\mu \leq n_\mu + 1 = \mu+1$. Therefore $m_\mu = \mu+1$. By induction $n_\mu = \mu$ and $m_\mu = \mu+1$ for all $\mu = 0,\cdots,p$ if $x \in S^*\langle v,a\rangle$.

Conversely, take $x \in S\langle v,\mathfrak{a}\rangle$ and define n_μ and m_μ as before. Then $m_\mu \geq \mu+1$ and $0 \leq m_\mu - n_\mu \leq 1$ for all $\mu = 0,\cdots,p$. Assume $n_\mu = \mu$ for all $\mu = 0,\cdots,p$. Then $m_\mu = \mu+1$ for all $\mu = 0,\cdots,p$. Assume

$x \in \partial_j S\langle v, \mathfrak{a}\rangle$ for some $j \in Z[0,p]$. Then $j = n_j \geq j+1$ which is wrong. Hence $x \in S^*\langle v, \mathfrak{a}\rangle$; q.e.d.

LEMMA 2.3. *Take* $v \in F^n$ *and* $\mathfrak{a}$ *and* $\mathfrak{b}$ *in* $\mathfrak{S}(p,n)$ *with* $\mathfrak{a} \neq \mathfrak{b}$. *Then* $S^*\langle v, \mathfrak{a}\rangle \cap S^*\langle v, \mathfrak{b}\rangle = \emptyset$.

Proof. A number $q \in Z[0,p]$ exists such that $a_\mu = b_\mu$ for $0 \leq \mu < q$ and $a_q \neq b_q$. W.l.o.g. $a_q < b_q$. Then $\hat{a}_q \leq \hat{b}_q - 1$. Assume $x \in S^*\langle v, \mathfrak{a}\rangle \cap S^*\langle v, \mathfrak{b}\rangle$ exists. Then

$$q+1 \leq \dim E(x) \cap E(v_{\hat{a}_q}) \leq \dim E(x) \cap E(v_{\hat{b}_q - 1}) = q$$

which is impossible; q.e.d.

THEOREM 2.4. *Take* $v \in F^n$ *and* $\mathfrak{a} \in \mathfrak{S}(p,n)$. *Then* $S\langle v, \mathfrak{a}\rangle$ *is an* $\vec{\mathfrak{a}}$*-dimensional irreducible analytic subset of* $G_p(V)$. *The open Schubert cell* $S^*\langle v, \mathfrak{a}\rangle$ *is biholomorphically equivalent to* $\mathbb{C}^{\vec{\mathfrak{a}}}$ *and is dense in* $S\langle v, \mathfrak{a}\rangle$.

The proof is long and complicated and has therefore been put into the appendix. It follows Chern [7] §8. Theorem 2.4 has important consequences for Schubert varieties and these will be discussed later.

THEOREM 2.5. *Take* $v \in F^n$. *Then* $G_p(V)$ *can be represented as a disjoint union of open Schubert cells*

$$G_p(V) = \bigcup_{\mathfrak{a} \in \mathfrak{S}(p,n)} S^*\langle v, \mathfrak{a}\rangle .$$

Proof. Take $x \in G_p(V)$. For $\mu \in Z[0,p]$ there exists a smallest integer $b_\mu \in Z[0,n]$ such that $\dim E(x) \cap E(v_{b_\mu}) \geq \mu+1$. Also define $m_\mu = \dim E(x) \cap E(v_{b_\mu})$ and $n_\mu = \dim E(x) \cap E(v_{b_\mu - 1})$. Then $0 \leq m_\mu - n_\mu \leq 1$. By definition of b_μ we have $n_\mu \leq \mu$. Hence $m_\mu \leq n_\mu + 1 \leq \mu + 1$. Therefore $m_\mu = \mu+1$ and $n_\mu = \mu$ for $\mu = 0,1,\cdots,p$. Assume that $b_\mu \leq b_{\mu-1}$ for $\mu \geq 1$. Then $E(v_{b_\mu}) \subseteq E(v_{b_{\mu-1}})$. Therefore $\mu+1 \leq m_\mu \leq m_{\mu-1} = \mu$, which is impossible. Therefore $0 \leq b_0 < b_1 < \cdots < b_p \leq n$. Define $a_\mu = b_\mu - \mu$. Then $0 \leq a_0 \leq a_1 \leq \cdots \leq a_p \leq n-p$ and $\mathfrak{a} \in \mathfrak{S}(p,n)$. Since $m_\mu \geq \mu+1$ for $\mu = 0,\cdots,p$, we have $x \in S\langle v,\mathfrak{a}\rangle$, and because $n_\mu = \mu$ for $\mu = 0,\cdots,p$, we have $x \in S^*\langle v,\mathfrak{a}\rangle$ by Lemma 2.3; q.e.d.

Of course this subdivision of the Grassmann manifold into open Schubert cells is well known and in fact is a CW-complex. For a proof see Milnor-Stasheff [23] Theorem 6.4. This subdivision implies that $G_p(V)$ is simply connected and that each $S\langle v,\mathfrak{a}\rangle$ is a cycle. There is no torsion in $H_m(G_p(V), Z)$ and in fact $H_m(G_p(V), Z) = 0$ if m is odd. Moreover $H_{2m}(G_p(V), C)$ is a complex vector space of dimension $\#\mathfrak{S}(p,n,m)$, where $\mathfrak{S}(p,n,m)$ is defined as in §1.b. In fact the cells $S^*\langle v,\mathfrak{a}\rangle$ (or $S\langle v,\mathfrak{a}\rangle$) with $\mathfrak{a} \in \mathfrak{S}(p,n,m)$ (and with $v \in F^n$ fixed) can be taken as a base of $H_{2m}(G_p(V), C)$. By duality $H^m(G_p(V), C) = 0$ if m is odd and $H^{2m}(G_p(V), C)$ has dimension $\#\mathfrak{S}(p,n,m)$. By Poincaré duality $H^{2m}(G_p(V), C)$ has also dimension $\#\mathfrak{S}(p,n,q)$ with $q = d(p,n)-m$. If $\mathfrak{a} \in \mathfrak{S}(p,n)$, define $\phi(\mathfrak{a}) = \mathfrak{a}^*$. Then

$$\phi : \mathfrak{S}(p,n,m) \to \mathfrak{S}(p,n,d(p,n)-m) \tag{2.11}$$

is bijective with $\phi^{-1} = \phi$, which is in agreement with the previous statement on Betti numbers. Since $G_p(V)$ is a complex symmetric space, the vector space $\mathrm{Inv}^{2m}(G_p(V)$ of invariant forms is isomorphic to $H^{2m}(G_p(V), C)$. Hence

$$\dim \mathrm{Inv}^{2m}(G_p(V)) = \#\mathfrak{S}(p,n,m) . \tag{2.12}$$

Now, we shall return to Schubert varieties and the consequences of Theorem 2.4.

THEOREM 2.6. *Take* $\mathfrak{a} \in \mathfrak{S}(p,n)$ *and* $v \in F(\mathfrak{a})$. *Then*:

1) *The Schubert variety* $S(v, \mathfrak{a})$ *is an irreducible analytic set of dimension* $\vec{\mathfrak{a}}$.

2) *The projection* $\sigma : S(\mathfrak{a}) \to F(\mathfrak{a})$ *has irreducible fibers of dimension* $\vec{\mathfrak{a}}$.

3) *The Schubert family* $S(\mathfrak{a})$ *is irreducible and has dimension* $d(\mathfrak{a}) + \vec{\mathfrak{a}}$. (*For the definition of* $d(\mathfrak{a})$ *see* §1b).

4) *The projection* $\pi : S(\mathfrak{a}) \to G_p(V)$ *has irreducible fibers of dimension* $d(\mathfrak{a}) + \vec{\mathfrak{a}} - d(p,n) = d(\mathfrak{a}) - \vec{\mathfrak{a}}^*$.

5) *If* $x \in G_p(V)$, *then* $S_x(\mathfrak{a})$ *is irreducible and has dimension* $d(\mathfrak{a}) + \vec{\mathfrak{a}} - d(p,n)$.

Proof. 1) A flag $w \in F^n$ exists such that $\phi_{\mathfrak{a}}(w) = v$. Hence $S(v, \mathfrak{a}) = S\langle w, \mathfrak{a}\rangle$ is irreducible and has dimension $\vec{\mathfrak{a}}$ by Theorem 2.4. Now 1) implies 2) trivially. Now 2) and Lemma 1.2 imply 3). Because $G_p(V)$ is simply connected, 3) and Lemma 1.3 imply 4) which is equivalent to 5); q.e.d.

3. CHERN FORMS

a) *Chern forms of vector bundles*

Let M be a complex space. Let $\pi : W \to M$ be a holomorphic vector bundle of fiber dimension k with a hermitian metric κ along the fibers of W. Let $e_1, \cdots, e_k$ be holomorphic sections of W over an open subset U of M such that $e_1(x), \cdots, e_k(x)$ is a base of the fiber W_x for each $x \in U$. Then $e = (e_1, \cdots, e_k)$ is called a *holomorphic frame* of W over U. For μ and ν in $\mathbb{Z}[1,k]$, the function $\kappa_{\mu\nu} = \kappa(e_\mu, e_\nu)$ is of class C^∞. A hermitian matrix $H_e = \overline{H}_e^t = \text{matrix}\,(\kappa_{\mu\nu})$ is defined with $\det H_e > 0$. The *connection matrix* $\theta_e = (\partial H_e) H_e^{-1}$ is a matrix of forms of bidegree $(1,0)$ and class C^∞ on U and the curvature matrix

$$\Omega_e = d\theta_e - \theta_e \wedge \theta_e = \bar{\partial}\theta_e \tag{3.1}$$

is a matrix of forms of bidegree (1.1). Note: Differential forms on complex spaces were introduced by Bloom-Herrera [2]. Tung [33] gives an excellent introduction to this topic complete with definitions and proofs. Other accounts, without proof, can be found in [19], [22], and [30]. Let $\tilde{e}$ be another holomorphic frame of W over an open set $\tilde{U}$ with $U \cap \tilde{U} \neq \emptyset$. Then $\tilde{e}^t = B e^t$, where B is a matrix of holomorphic functions on $U \cap \tilde{U}$ with $\det B(x) \neq 0$ for all $x \in U \cap \tilde{U}$. Let I_k be the unit matrix of k-lines. Then

$$I_k + \frac{i}{2\pi}\Omega_{\tilde{e}} = B(I_k + \frac{i}{2\pi}\Omega_e)B^{-1}$$

on $U \cap \tilde{U}$. Therefore one and only one form $c(W, \kappa)$ of class C^∞ exists on M such that

$$c(W,\kappa) \,|\, U = \det(I_k + \frac{i}{2\pi}\Omega_e) \tag{3.2}$$

for every holomorphic frame e of W. Here $c(W,\kappa)$ is called the *total Chern form* of W for κ. For each $q \in \mathbb{Z}[0,k]$ there exists a unique form $c_q(W,\kappa)$ of bidegree (q,q) and class C^∞ such that

$$(3.3) \qquad c(W,\kappa) = \sum_{q=0}^{k} c_q(W,\kappa) .$$

Here $c_q(W,\kappa)$ is called the qth *Chern form* of W for κ. Define $c_q(W,\kappa) = 0$ if $q < 0$ or if $q > k$. Observe that $\partial c_q(W,\kappa) = 0$ and $\bar{\partial} c_q(W,\kappa) = 0$ for all $q \in \mathbb{Z}$.

If W is trivial, a form χ of class C^∞ exists such that $c(W,\kappa) = dd^c\chi$. If κ_1 and κ_2 are hermitian metrics along the fibers of W, a form ξ of class C^∞ exists such that

$$(3.4) \qquad c(W,\kappa_1) = c(W,\kappa_2) + dd^c\xi .$$

Let N be a complex space. Let $f: N \to M$ be a holomorphic map. Let $(\widetilde{W},\widetilde{\kappa})$ be the pull back of (W,κ). Then

$$(3.5) \qquad f^*(c(W,\kappa)) = c(\widetilde{W},\widetilde{\kappa}) .$$

Because $f(N)$ may be contained in the singularities of M, the existence and the properties of the pull back operator f^* are non-trivial. See Tung [33]. If W^* is the dual bundle to W and if κ^* is dual to κ, then

$$(3.6) \qquad c_q(W^*,\kappa^*) = (-1)^q c_q(W,\kappa) \qquad \forall\, q \in \mathbb{Z} .$$

If W_λ is a holomorphic vector bundle with hermitian metric κ_λ for $\lambda = 1, 2$, then

$$(3.7) \qquad c(W_1 \oplus W_2, \kappa_1 \oplus \kappa_2) = c(W_1,\kappa_1) \wedge c(W_2,\kappa_2) .$$

If W_λ is a holomorphic vector bundle with hermitian metric κ_λ for $\lambda = 1,2,3$, and if $0 \to W_1 \to W_2 \to W_3 \to 0$ is a short exact sequence, then there exists a form ξ of class C^∞ such that

$$c(W_1, \kappa_1) \wedge c(W_3, \kappa_3) = c(W_2, \kappa_2) + dd^c \xi \ . \tag{3.8}$$

The Chern forms of tensor products and exterior products are more difficult to compute. A formal calculus can be used. First let $\mathfrak{T}(p,q)$ be the set of all increasing, injective maps $\mu : \mathbf{N}[1,p] \to \mathbf{N}[1,q]$. Formally write

$$c(W, \kappa) = \prod_{j=1}^{k} (1 + \lambda_j) \ . \tag{3.9}$$

Then

$$c(W^*, \kappa^*) = \prod_{j=1}^{k} (1 - \lambda_j) \tag{3.10}$$

$$c(\bigwedge_p W, \bigwedge_p \kappa) = \prod_{\mu \in \mathfrak{T}(p,k)} (1 + \lambda_{\mu(k)} + \cdots + \lambda_{\mu(k)}) \ . \tag{3.11}$$

If $p = k$ is the fiber dimension of W , this implies

$$c_1(\bigwedge_k W, \bigwedge_k \kappa) = c_1(W, \kappa) \ . \tag{3.12}$$

If W_ρ is a holomorphic vector bundle with hermitian metric κ_ρ for $\rho = 1,2$, write formally

$$c(W_\rho, \kappa_\rho) = \prod_{j=1}^{k_\rho} (1 + \lambda_j^\rho) \ . \tag{3.13}$$

Then

$$c(W_1 \oplus W_2, \kappa_1 \oplus \kappa_2) = \prod_{j=1}^{k_1} (1 + \lambda_j^1) \prod_{j=1}^{k_2} (1 + \lambda_j^2) \tag{3.14}$$

$$c(W_1 \oplus W_2, \kappa_1 \oplus \kappa_2) = \prod_{j=1}^{k_1} \prod_{i=1}^{k_2} (1 + \lambda_j^1 + \lambda_i^2) \ . \tag{3.15}$$

The products are to be expressed as polynomials in the elementary symmetric functions of $\lambda_1, \cdots, \lambda_k$, respectively of $\lambda_1^\rho, \cdots, \lambda_{k_\rho}^\rho$ which are Chern forms.

Assume $\dim M < \infty$. Let $A(M)$ be the vector space of forms of class C^∞ on M. Let $A^p(M)$ be the linear subspace of forms of degree p. Define $Z(M) = \{\omega \in A(M) \mid d\omega = 0\}$ and $Z^p(M) = A^p(M) \cap Z(M)$. Define $B(M) = d\,A(M)$ and $B^p(M) = B(M) \cap A^p(M)$ with $B^0(M) = 0$. The quotient spaces $R(M) = Z(M)/B(M)$ and $R^p(M) = Z^p(M)/B^p(M)$ are called the *de Rham groups*. Here $R(M)$ is the direct sum of the $R^p(M)$ for $0 \leq p \in \mathbb{Z}$ and $R(M)$ is an exterior graded algebra. Let ρ be the residual map. The *Chern classes* of a holomorphic vector bundle W of fiber dimension k are defined by

$$c_p(W) = \rho(c_p(W,\kappa)) \in R^p(M)$$

$$c(W) = \rho(c(W,\kappa)) = \sum_{p=0}^{k} c_p(W,\kappa) \in R(M) .$$

They do not depend on the hermitian metric κ chosen. The computational rules carry over to the Chern classes. In particular, if $0 \to W_1 \to W_2 \to W_3 \to 0$ is a short exact sequence, then

$$c(W_1) \wedge c(W_3) = c(W_2) . \tag{3.16}$$

If M is a complex manifold, then $R(M) = H^*(M, \mathbb{C})$ and $R^p(M) = H^p(M, \mathbb{C})$ can be identified with the cohomology groups by the de Rham isomorphism. Then the Chern classes are cohomology classes.

For this section see Bott-Chern [3], Chern [7], Hirzebruch [16] and standard textbooks.

b) *The basic Chern forms*

Let V be a complex vector space of dimension $n+1$. Take $p \in \mathbb{Z}[0,n]$. Over $G_p(V)$, the short exact sequence

$$(3.17) \qquad 0 \longrightarrow S_p(V) \xrightarrow[j]{} G_p(V) \times V \xrightarrow[\eta]{} Q_p(V) \longrightarrow 0$$

is defined and is called the *classification sequence*. Take a positive definite hermitian form $\mathfrak{l}$ on V. Then $\mathfrak{l}$ defines hermitian metrics along the fibers of $G_p(V) \times V$ and along the fibers of the sub-bundle $S_p(V)$. Its complementary bundle $S_p(V)^\perp$ in $G_p(V) \times V$ is differentiably isomorphic to $Q_p(V)$. Then $\mathfrak{l}$ restricts to $S_p(V)^\perp$ and carries over to a hermitian metric $\mathfrak{l}$ along the fibers of $Q_p(V)$ under the isomorphism $\eta : S_p(V)^\perp \to Q_p(V)$. Define the *basic Chern forms* by

$$(3.18) \qquad s_q[p] = c_q(S_p(V), \mathfrak{l})$$

$$(3.19) \qquad s[p] = c(S_p(V), \mathfrak{l}) = \sum_{q=0}^{p+1} s_q[p]$$

$$(3.20) \qquad c_q[p] = c_q(Q_p(V), \mathfrak{l})$$

$$(3.21) \qquad c[p] = c(Q_p(V), \mathfrak{l}) = \sum_{q=0}^{n-p} c_q[p] .$$

The hermitian form $\mathfrak{l}$ on V defines an unitary group $\mathfrak{U} = \mathfrak{U}(V, \mathfrak{l})$. Since (3.17) is invariant under $\mathfrak{U}(V)$, the Chern forms defined in (3.18)-(3.21) are invariant under the action of $\mathfrak{U}$ and can be identified with their cohomology classes by the de Rham isomorphism. Define $c_q[p] = 0$ if $q < 0$ or $q > n-p$ and $s_q[p] = 0$ if $q < 0$ or $q > p+1$. Then (3.8) and (3.16) imply that $c[p] \wedge s[p] = 1$. Trivially $c_0[p] = s_0[p] = 1$. Hence

$$(3.22) \qquad \sum_{\mu+\nu=q} c_\mu[p] \vee s_\nu[p] = 0 \qquad \text{if } q \in \mathbb{N} .$$

Therefore the $s_\nu[p]$ can be computed recursively by the $c_\mu[p]$, and the $c_\mu[p]$ by the $s_\nu[p]$.

The name classifying sequence can be explained as follows: Let W be a holomorphic vector bundle of fiber dimension k over the complex space M. Let V be a complex vector space of dimension $n+1$ with $p = n-k \geq 0$. Let $\eta : M \times V \to W$ be a surjective holomorphic vector bundle homorphism. Then η is called an *amplification* and W is called *ample*. Let S be the kernel of η. A short exact sequence

$$0 \longrightarrow S \xrightarrow[j]{} M \times V \xrightarrow[\eta]{} W \longrightarrow 0 \tag{3.22}$$

is defined. Here S has fiber dimension $p+1$. For each $x \in M$, a point $\phi(x) \in G_p(V)$ is defined such that $S_x = E(\phi(x))$. The classification map $\phi : M \to G_p(V)$ so defined is holomorphic and pulls (3.17) back to (3.23). Take a positive definite hermitian form $\mathfrak{l}$ on V. Then ℓ defines hermitian metrics $\mathfrak{l}$ along the fibers of $G_p(V) \times V, S_p(V)$ and $G_p(V)$. The same method defines hermitian metrics $\mathfrak{l}$ along the fibers of $M \times V$, S and W. Then

$$c(W, \mathfrak{l}) = \phi^*(c[p]) \qquad c(S, \mathfrak{l}) = \phi^*(s[p]) . \tag{3.24}$$

In this sense the Chern forms $c[p]$ and $s[p]$ are universal. Observe that a holomorphic vector bundle on a Stein manifold is ample.

Again let V be a complex vector space of dimension $n+1 > 0$. Let ℓ be a positive definite hermitian form on V. Then $\mathfrak{l}$ defines a Fubini-Study Kaehler metric on $P(V)$ and on each $G_p(V)$ whose exterior forms are denoted by ω respectively ω_p.

PROPOSITION 3.1. *If* $q \in Z[0,n]$, *then* $s_1[0] = -\omega$ *and* $c_q[0] = \omega^q$.

Proof. Take $0 \neq \alpha \in V^*$. Let $E[\alpha]$ be the kernel of α and define $\ddot{E}[\alpha] = P(E[\alpha])$. A holomorphic frame v_α of $S_0(V)$ over $P(V) - \ddot{E}[\alpha]$ is defined by $v_\alpha(x) = (x, \mathfrak{x}/\alpha(\mathfrak{x}))$ where $x \in P(V) - \ddot{E}[\alpha]$ and $0 \neq \mathfrak{x} \in V$ with $P(\mathfrak{x}) = x$. By the definition of $\mathfrak{l}$, we have

$$|v_\alpha(x)|_{\mathfrak{l}} = \frac{|\mathfrak{x}|}{|\alpha(\mathfrak{x})|} .$$

Then $s_1[0] = -dd^c \log |v_\alpha|_{\mathfrak{l}}^2$ on $\mathbf{P}(V) - \ddot{E}[\alpha]$. Define $\tau : V \to \mathbf{R}_+$ by $\tau(\mathfrak{x}) = |\mathfrak{x}|^2$. Then

$$\begin{aligned} \mathbf{P}^*(s_1[0]) &= -dd^c \log |v_\alpha \circ \mathbf{P}|_{\mathfrak{l}}^2 = dd^c \log(|\alpha|^2/\tau) \\ &= dd^c \log \tau = -\mathbf{P}^*(\omega) . \end{aligned}$$

Therefore $s_1[0] = -\omega$. Observe that $c_0[0] = 1 = \omega^0$. Assume that $c_q[0] = \omega^q$ has already been proved. Now (3.22) implies that

$$c_{q+1}[0] + c_q[0] \wedge s_1[0] = 0 .$$

Hence $c_{q+1}[0] = c_q[0] \wedge \omega = \omega^{q+1}$; q.e.d.

PROPOSITION 3.2. $c_1[p] = \omega_p$.

Proof. Take $x \in G_p(V)$. Then $x = \mathbf{P}(\mathfrak{x}_0 \wedge \cdots \wedge \mathfrak{x}_p)$ where $\mathfrak{x}_q \in V$. Then the following identification can be made:

$$S_p(V)_x = E(x) = \mathbf{C}\mathfrak{x}_0 + \cdots + \mathbf{C}\mathfrak{x}_p$$

$$(\textstyle\bigwedge_{p+1} S_p(V))_x = \textstyle\bigwedge_{p+1} E(x) = \mathbf{C}\mathfrak{x}_0 \wedge \cdots \wedge \mathfrak{x}_p$$

$$S_0(\textstyle\bigwedge_{p+1} V)_x = E(x, \textstyle\bigwedge_{p+1} V) = \mathbf{C}\mathfrak{x}_0 \wedge \cdots \wedge \mathfrak{x}_p = (\textstyle\bigwedge_{p+1} S_p(V))_x$$

$$\textstyle\bigwedge_{p+1} S_p(V) = S_0(\textstyle\bigwedge_{p+1} V) \mid G_p(V) .$$

The induced hermitian metrics agree. Proposition 3.1 implies that

$$c_1(\textstyle\bigwedge_{p+1} S_p(V), \mathfrak{l}) = c_1(S_0(\textstyle\bigwedge_{p+1} V), \mathfrak{l}) = -\omega_p$$

on $G_p(V)$. Now (3.12) implies that $s_1[p] = c_1(\bigwedge_{p+1} S_p(V), I) = -\omega_p$. Hence $c_1[p] = -s_1[p] = \omega_p$; q.e.d.

For this section compare Chern [7] and Matsushima-Stoll [22].

4. THE THEOREM OF BOTT AND CHERN

a) *Remarks on fiber integration*

Through the remainder of this monograph fiber integration will be used extensively. If the fibers are smooth and the map is regular, the fiber integration operator is well known. An exposition can be found [27] Appendix II and in [28]. However, difficulties appear, if the fibers are not smooth, since the fiber integration depends on the splitting of the cotangent bundle by smooth fibers. Nevertheless, fiber integration exists for projective holomorphic maps (definition below) and was used first by Cowen [8]. A solid foundation was created by Tung [33], where all the pertinent properties of [27] Appendix II were proved for projective maps between complex spaces.

Let M and N be complex spaces of pure dimensions m and n respectively, with $m-n=q\geq 0$. A holomorphic map $f:M\to N$ is said to be *projective* at $a\in M$ if there exist open neighborhoods U of a in M and V of $f(a)$ in N, a pure q-dimensional complex space W and a biholomorphic map $\alpha:U\to V\times W$ such that $f|U=\pi\circ\alpha$ where $\pi:V\times W\to V$ is the projection. The map f is said to be *locally trivial* over $b=f(a)$ if $U=f^{-1}(V)$. The map f is said to be *projective* (*respectively locally trivial*) if f is projective at all $a\in M$ (respectively locally trivial over all $b\in N$). For projective holomorphic maps Tung [33] VIII §8.2 defines the fiber integration operator and proves the theorems of [27] Appendix II. He does not prove the exterior product theorem of [22] Theorem 1.2, but the same proof also works for projective maps between complex spaces. Here Tung [33] 8.2.9 is used. In fact, the theorem states

THEOREM 4.1. *Let* X, Y *and* N *complex spaces of pure dimensions* k,

m *and* n *respectively. Assume* $p = k - n \geq 0$ *and* $q = m - n \geq 0$ *. Let* $f : X \to N$ *and* $g : Y \to N$ *be proper, projective, surjective holomorphic maps. Let*

$$Z = \{(x,y) \in X \times Y \mid f(x) = g(y)\}$$

be the relative product of f *and* g *. Let* $\pi_1 : Z \to X$ *and* $\pi_2 : Z \to Y$ *be the projections. Define* $h : Z \to N$ *by* $h = f \circ \pi_1 = g \circ \pi_2$ *. Let* ψ *be a continuous form of degree* r *on* X *with* $r \geq 2p$ *and let* χ *be a continuous form of degree* s *on* Y *with* $s \geq 2q$ *. Then* π_1*,* π_2 *and* h *are proper, projective, surjective holomorphic maps and*

$$h_*(\pi_1^*(\psi) \wedge \pi_2^*(\chi)) = f_*(\psi) \wedge g_*(\chi) .$$

Of course fiber integration can be defined as a current for any holomorphic map. This is trivial. But the fiber integration current is a form for projective maps, and this is not trivial. However, in this section fiber integration will be used for regular maps only. In the subsequent sections fiber integration will be used for locally trivial maps with fibers containing singularities, and Theorem 4.1 will play an important role.

b) *Representation and integration of Chern forms*

THEOREM 4.2 (Bott-Chern [3]). *Let* V *be a complex vector space of dimension* $n+1 \geq 1$ *. Let* l *be a positive definite hermitian form on* V *. Consider the short flag* F_p *with the projections* $\tau : F_p \to G_p(V)$ *and* $\pi : F_p \to P(V)$ *. If* $0 \leq q \leq n-p$ *, then*

$$c_q[p] = \tau_* \pi^*(\omega^{p+q}) \geq 0 . \tag{4.1}$$

Proof. If the classification sequences on $P(V)$ and $G_p(V)$ are lifted to F_p , a commutative diagram (4.2) of exact sequences is obtained. Here (j_0, η_0) and (j_p, η_p) are the pull back exact sequences. If $(x,y) \in F_p$, then $E(y) \subseteq E(x)$. Therefore

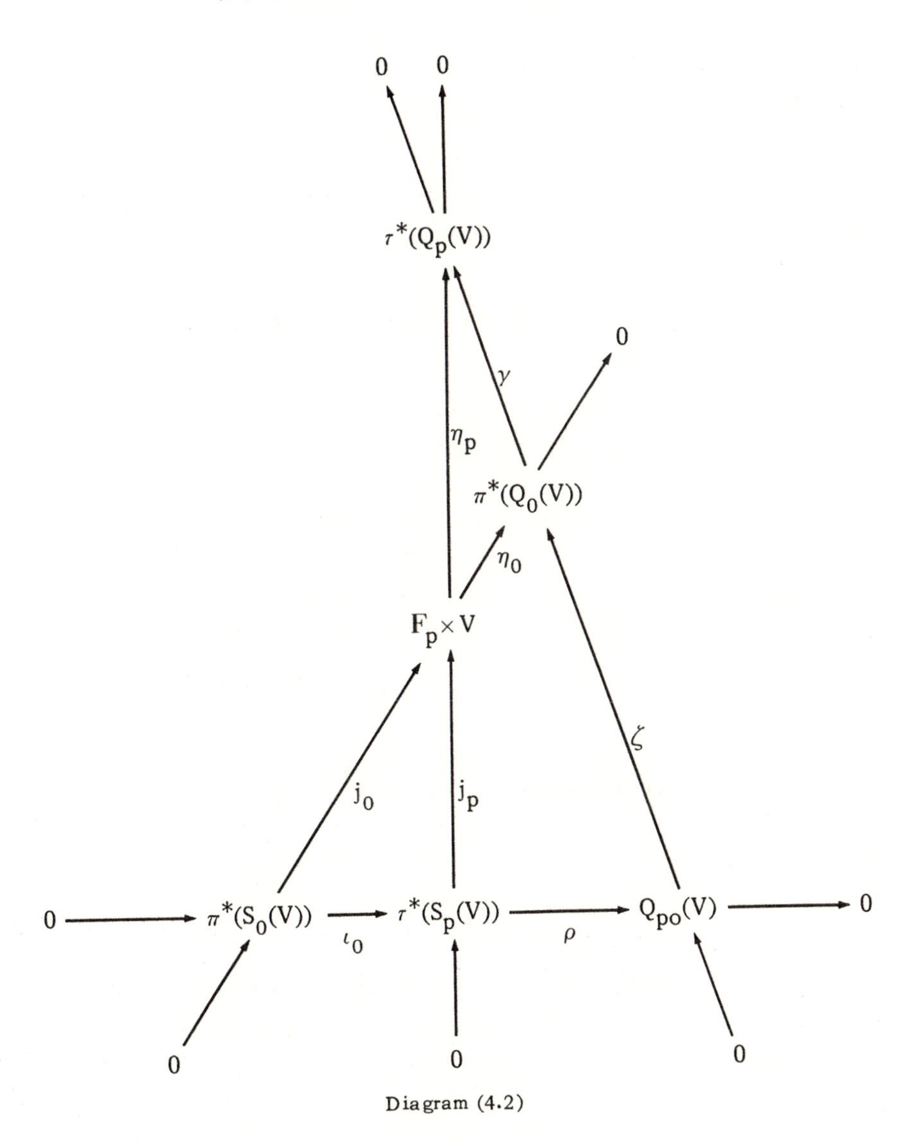

Diagram (4.2)

$$\pi^*(S_0(V)) = \{(x,y,\mathfrak{v}) \in F_p \times V \mid \mathfrak{v} \in E(y)\}$$

$$\subseteq \{(x,y,\mathfrak{v}) \in F_p \times V \mid \mathfrak{v} \in E(x)\} = \tau^*(S_p(V)) .$$

The inclusion ι_0 is defined. The exact sequences (j_p, η_p) and (j_0, η_0) and the inclusion ι_0 define the epimorphism γ. Let $Q_{po}(V)$ be the quotient bundle of $\tau^*(S_p(V))$ over $\pi^*(S_0(V))$ and let ρ be the residual map. A monomorphism ζ is uniquely defined by $\zeta \circ \rho = \eta_0 \circ j_p$. Then (ζ, γ) is exact and the diagram commutes.

Hermitian metrics $\mathfrak{l}$ on $F_p \times V$ and by restrictions on $\pi^*(S_0(V))$ and $\tau^*(S_p(V))$ are defined. By orthogonality, quotient metrics along the fibers of $Q_{p0}(V)$, $\pi^*(Q_0(V))$ and $\tau^*(Q_p(V))$ are defined. In the sequences (j_p, η_p) and (j_0, η_0) these metrics are the pull back metrics. Define $d_q = c_q(Q_{p0}(V), \mathfrak{l})$ and $d = c(Q_{p0}(V), \mathfrak{l})$. Then $d = d_0 + \cdots + d_p$. By (3.8) there is a form $\xi = \xi_1 + \cdots + \xi_n$ of class C^∞ on F_p such that ξ_q has bidegree $(q-1, q-1)$ and such that

$$d \wedge \tau^*(c[p]) = \pi^*(c[0]) + dd^c \xi$$

$$\begin{aligned} \pi^*(\omega^{p+q}) &= \pi^*(c_{p+q}[0]) \\ &= d_p \wedge \tau^*(c_q[p]) + \sum_{\lambda=1}^{p} d_{p-\lambda} \wedge \tau^*(c_{q+\lambda}[p]) - dd^c \xi_{p+q} . \end{aligned}$$

Because τ has fiber dimension p, we obtain

$$\tau_* \pi^*(\omega^{p+q}) = \tau_*(d_q) c_q[p] - dd^c \tau_*(\xi_{p+q}) . \tag{4.3}$$

Now, $\tau_*(d_q)$ has to be computed. Take $x \in G_p(V)$. Then $\tau^{-1}(x) = \{x\} \times \ddot{E}(x)$ is identified with $\ddot{E}(x)$. Let $j_x : \ddot{E}(x) \to P(V)$ be the inclusion. The restriction of the hermitian metric $\mathfrak{l}$ to $E(x)$ gives a Fubini-Study Kaehler metric on $\ddot{E}(x)$ whose exterior form is $j_x^*(\omega)$. Therefore

$$c_p(Q_0(\ddot{E}(x)), \mathfrak{l}) = j_x^*(\omega^p) .$$

Let $\tilde{j}_x : \ddot{E}(x) \to F_p$ be the inclusion map. Then

$$\tilde{j}_x^* \pi^*(S_0(V)) = \bigcup_{y \in \ddot{E}(x)} \{y\} \times E(y) = S_0(\ddot{E}(x))$$

because $E(y) \subseteq E(x)$. Also

$$\tilde{j}_x^* \tau^*(S_p(V)) = \bigcup_{y \in \ddot{E}(x)} \{y\} \times E(x) = \ddot{E}(x) \times E(x) .$$

Hence $\tilde{j}_x : \ddot{E}(x) \to F_p$ pulls back the exact sequence (ι_0, ρ) to

$$0 \to S_0(\ddot{E}(x)) \to \ddot{E}(x) \times E(x) \to Q_0(\ddot{E}(x)) \to 0$$

and the pull back hermitian metrics are consistent with the hermitian metrics introduced by $I | E(x)$. Therefore

$$\tilde{j}_x^*(d_p) = c_p(Q_0(\ddot{E}(x)), I) = j_x^*(\omega^p)$$

which implies that

$$\tau_*(d_p) = \int_{\ddot{E}(x)} \tilde{j}_x^*(d_p) = \int_{\ddot{E}(x)} j_x^*(\omega^p) = 1 .$$

Now, (4.3) shows that $dd^c \tau_*(\xi_{p+q})$ is invariant under the actions of the unitary group on $G_p(V)$ which is a symmetric space. Therefore $dd^c \tau_*(\xi_{p+q}) = 0$ and (4.1) is proved. Indeed, $c_q[p] \geq 0$ since $\omega^{p+q} > 0$; q.e.d.

For the remainder of this section, V is a complex vector space of dimension $n+1 \geq 1$ and I is a positive definite hermitian form on V. The Theorem of Bott and Chern shall be extended to flag spaces F_{pq}. Several preparations are needed.

LEMMA 4.3. *Let* p *and* q *be integers with* $0 \leq q < p \leq n$. *Consider the short flag space* F_{pq} *with the projections* $\pi : F_{pq} \to G_q(V)$ *and* $\tau : F_{pq} \to G_p(V)$. *Take* $y \in G_q(V)$. *Define* $W = V/E(y)$. *Introduce the quotient metric on* W. *Let* $\rho : V \to W$ *be the residual map. Define* $s = p-q-1$. *Then a smooth, injective holomorphic map* $j : G_s(W) \to G_p(V)$ *is defined by*

$$E(z) = \rho(E(j(z)) = E(j(z))/E(y)$$

for all $z \in G_s(W)$. *Define* $N = \tau(\pi^{-1}(y))$. *Then* $j : G_s(W) \to N$ *is biholomorphic. The map* j *pulls back* $Q_p(V)$ *isometrically to* $Q_s(W)$ *such that* $j^*(c_\mu[p,V]) = c_\mu[s,W]$ *for* $\mu = 0, 1, \cdots, n-p$.

Proof. By Lemma 1.3, π is surjective and locally trivial. Because $\pi \circ g = g \circ \pi$ for all $g \in GL(V)$, the map π is regular. Hence $\pi^{-1}(y)$ is a smooth, compact complex submanifold of F_{pq}. Since $\pi^{-1}(y) = N \times \{y\}$, also N is a smooth, compact, complex submanifold of $G_p(V)$ and $\tau : \pi^{-1}(y) \to N$ is biholomorphic. Observe $N = \{x \in G_p(V) \mid E(y) \subseteq E(x)\}$. Identify $W = E(y)^\perp$. Then $\rho : V \to W$ is the projection and the hermitian metric on V is obtained by restriction from V. Take $z \in G_s(W)$. Then $E(z) \subseteq W \subseteq V$ and $E(z) \cap E(y) = 0$. Therefore there exists a unique $j(z) \in G_p(V)$ such that $E(j(z)) = E(y) + E(z)$. This sum is direct and $\rho(E(j(z))) = E(z)$.

Take $z_0 \in G_s(W)$. Holomorphic vector functions $\mathfrak{w}_\lambda : U \to W$ exist on an open neighborhood U of z_0 in $G_s(W)$ for $\lambda = 0, 1, \cdots, s$ such that $\mathfrak{w}(z) = \mathfrak{w}_0(z) \wedge \cdots \wedge \mathfrak{w}_s(z) \neq 0$ and $z = P(\mathfrak{w}(z))$ for all $z \in U$. Take $0 \neq \mathfrak{y} \in \widetilde{G}_q(V)$ with $y = P(\mathfrak{y})$. Observe that $W \subseteq V$. Hence $0 \neq \mathfrak{w}(z) \in \widetilde{G}_s(V)$ and $0 \neq \mathfrak{y} \wedge \mathfrak{w}(z) \in \widetilde{G}_p(V)$ with $j(z) = P(\mathfrak{y} \wedge \mathfrak{w}(z))$ for all $z \in U$. Therefore j is holomorphic.

Take $x \in N$. Then $E(y) \subseteq E(x)$. Therefore there exists $z \in G_s(V)$ such that $E(y) \cap E(z) = 0$, such that $E(y) + E(z) = E(x)$ and $E(z)$ is perpendicular to $E(y)$. Then $E(z) \subseteq W$, hence $z \in G_s(W)$. Therefore $x = j(z)$. The map j is surjective. Take z and z' in $G_s(W)$ with $j(z) = j(z')$. Then $E(z) = \rho(E(j(z)) = \rho(E(j(z'))) = E(z')$. Hence $z = z'$. The map j is injective. Because $G_s(W)$ and N are complex manifolds, the bijective holomorphic map $j : G_s(W) \to N$ is biholomorphic and $j : G_s(W) \to G_p(V)$ is smooth. In particular, N is connected and has dimension $(n-p)(p-q)$. Identify $N = G_s(W)$ such that j becomes the inclusion map. Because

$$0 \longrightarrow E(y) \longrightarrow E(j(z)) \xrightarrow{\rho} E(z) \longrightarrow 0$$

is exact, diagram 4.4 is defined. Its rows and columns are exact.

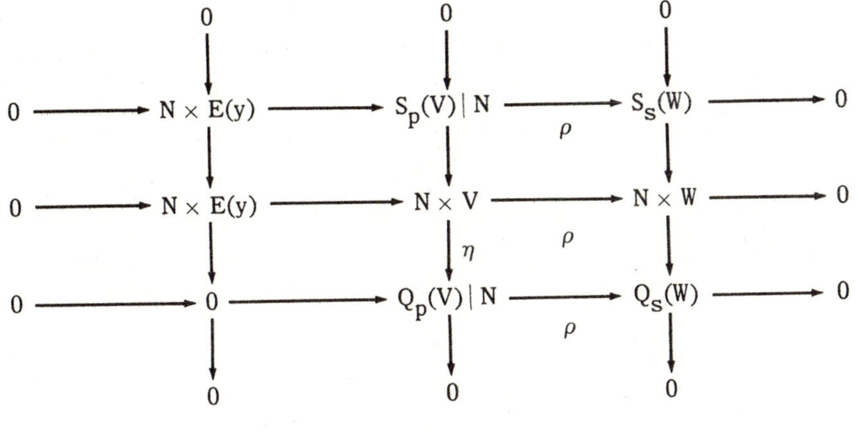

Diagram 4.4

Hence $\rho : Q_p(V)|N \to Q_s(W)$ is defined and is an isomorphism. Take $z \in N$. Let $H = E(z)^{\perp}$ be the orthogonal complement of $E(z)$ within W. By the definition of the hermitian metric along the fibers of $Q_s(W)$, the restriction $\eta_z = \eta : H \to Q_s(W)_z$ is an isomorphism. The metric on V restricts to W which restricts to H. Hence $V = E(y) + E(z) + H$ is an orthogonal decomposition with $E(j(z)) = E(y) + E(z)$. Therefore $H = E(j(z))^{\perp}$ in V. By the definition of the hermitian metric along the fibers of $Q_p(V)$ the restriction $\eta_{j(z)} = \eta : H \to Q_p(V)_{j(z)}$ is an isometry. Hence $\rho_z = \rho : Q_p(V)_{j(z)} \to Q_s(W)_z$ is an isometry since $\rho_z \circ \eta_{j(z)} = \eta_z$. Therefore $\rho : Q_p(V)|N \to Q_s(W)$ is an isometry. If $z \in G_s(W)$, then $E(z) = \rho(E(j(z)))$. Hence $Q_s(W)$ is the pull pack of $Q_p(V)$ under j, where the induced map $\rho^{-1} : Q_s(W) \to Q_p(V)|N$ is an isometry. Hence $j^*(c_\mu[p,V]) = c_\mu[s,W]$ for $\mu = 0, 1, \cdots, p-1$; q.e.d.

This result permits the computation of a number of integrals.

THEOREM 4.4. $$\int_{G_p(V)} c_{n-p}[p]^{p+1} = 1 .$$

Proof. If $p = 0$, then $G_p(V) = \mathbf{P}(V)$ and $c_n[0] = \omega^n$. The theorem is correct in this case. Now, the result shall be proven for $p \geq 1$, provided it is correct for $p-1$. Consider the flag space $\mathbf{F}_p$ with the projections $\pi : \mathbf{F}_p \to \mathbf{P}(V)$ and $\tau : \mathbf{F}_p \to G_p(V)$. Take $y \in \mathbf{P}(V)$ and define $W = V/E(y)$. Define $N = \tau\pi^{-1}(y)$. A biholomorphic map $j : G_{p-1}(W) \to N$ is defined by $E(z) = E(j(z))/E(y)$ if $z \in G_{p-1}(W)$. Induction and Theorem 4.3 imply that

$$\pi_*\tau^*(c_{n-p}[p]^p)(y) = \int_{\pi^{-1}(y)} \tau^*(c_{n-p}[p]^p)$$

$$= \int_N c_{n-p}[p,V]^p$$

$$= \int_{G_{p-1}(W)} c_{n-p}[p-1,W]^p = 1$$

$$\int_{G_p(V)} c_{n-p}[p]^{p+1} = \int_{G_p(V)} c_{n-p}[p]^p \wedge \tau_*\pi^*(\omega^n)$$

$$= \int_{\mathbf{F}_p} \tau^*(c_{n-p}[p]^p) \wedge \pi^*(\omega^n)$$

$$= \int_{\mathbf{P}(V)} \pi_*\tau^*(c_{n-p}[p]^p)\omega^n = 1 ;$$

q.e.d.

Consequently, $c_{n-p}[p]^{p+1} > 0$ at at least one point of $G_p(V)$. Because this form is invariant under the action of the unitary group on $G_p(V)$, we have $c_{n-p}[p]^{p+1} > 0$ at every point of $G_p(V)$.

PROPOSITION 4.5. *Let* p *and* q *be integers with* $0 \leq q < p \leq n$. *Let* $\pi : F_{pq} \to G_q(V)$ *and* $\tau : F_{pq} \to G_p(V)$ *be the projections. Then*

$$\pi_* \tau^* (c_{n-p}[p]^{p-q}) = 1 .$$

Proof. Define $s = p-q-1$. Take $y \in G_q(V)$. Define $W = V/E(y)$ and $N = \tau\pi^{-1}(y)$. A biholomorphic map $j : G_s(W) \to N$ is defined by $E(z) = E(j(z))/E(y)$ for $z \in G_s(W)$. Lemma 4.3 and Theorem 4.4 imply that

$$\pi_* \tau^* (c_{n-p}[p]^{p-q})(y) =$$

$$= \int_{\pi^{-1}(y)} \tau^*(c_{n-p}[p,V]^{p-q}) = \int_N c_{n-p}[p,V]^{p-q}$$

$$= \int_{G_s(W)} c_{n-p}[s,W]^{p-q} = 1 ;$$

q.e.d.

PROPOSITION 4.6. *Take* $p \in \mathbf{N}[1,n]$. *Let* $\pi : F_p \to \mathbf{P}(V)$ *and* $\tau : F_p \to G_p(V)$ *be the projections. Then*

$$\pi_* \tau^* (c_1[p]^{(n-p)p}) = D(p-1, n-1) .$$

Proof. Define $s = p-1$. Take $y \in \mathbf{P}(V)$. Define $W = V/E(y)$. Define $M = \pi^{-1}(y)$ and $N = \tau(\pi^{-1}(y))$. A biholomorphic map $j : G_s(W) \to N$ is defined by $E(z) = E(j(z))/E(y)$. Lemma 4.3 implies that

$$\pi_*\tau^*(c_1[p]^{(n-p)p})(y) =$$

$$= \int_M \tau^*(c_1[p,V]^{(n-p)p}) = \int_N c_1[p,V]^{(n-p)p}$$

$$= \int_{G_s(W)} c_1[s,W]^{(n-p)p} = \int_{G_s(W)} (\omega_s)^{(n-1-s)(s+1)} = D(s,n-1)\ ;$$

q.e.d.

PROPOSITION 4.7. *Take* $p \in \mathbf{N}[1,n]$. *Then*

$$\int_{G_p(V)} c_{n-p}[p] \wedge c_1[p]^{(n-p)p} = D(p-1,n-1)\ .$$

Proof. Let $\pi : \mathbf{F}_p \to \mathbf{P}(V)$ and $\tau : \mathbf{F}_p \to G_p(V)$ be the projections. Abbreviate $d = (n-p)p$. Then

$$\int_{G_p(V)} c_{n-p}[p] \wedge c_1[p]^d = \int_{G_p(V)} \tau_*\pi^*(\omega^n) \wedge c_1[p]^d$$

$$= \int_{\mathbf{F}_p} \pi^*(\omega^n) \wedge \tau^*(c_1[p])^d = \int_{\mathbf{P}(V)} \pi_*\tau^*(c_1[p]^d)\omega^n$$

$$= D(p-1,n-1)\ ;$$

q.e.d.

THEOREM 4.8 (Bott-Chern). *Let* p,q *and* μ *be integers with* $0 \le q \le p \le n$ *and* $p \le \mu \le n$. *Let* $\pi : \mathbf{F}_{pq} \to G_q(V)$ *and* $\tau : \mathbf{F}_{pq} \to G_p(V)$ *be the projections. Then*

$$D(q-1,p-1)\, c_{\mu-p}[p] = \tau_*\pi^*(c_{\mu-q}[q] \wedge c_1[q]^{(p-q)q})\ .$$

REMARK. The theorem was stated incorrectly by Bott and Chern [3] if $q > 0$. The correct version was stated in [28]. At last a proof shall be provided. For $q = 0$, see Theorem 4.2.

Proof. The flag space

$$F = \{(x,y,z) \in G_p(V) \times G_q(V) \times P(V) \mid E(z) \subseteq E(y) \subseteq E(x)\}$$

is a smooth, compact, analytic subset of $G_p(V) \times G_q(V) \times P(V)$. (In fact if $0 < q < p$, then F is isomorphic to $F(0, q-1, p-2)$ under the isomorphism $(x,y,z) \to (z,y,x)$.) The maps in diagram 4.5 are defined as projections. They are holomorphic and surjective and commute with the action of the unitary group. Hence they are locally trivial with smooth fibers whose dimension is the dimension of the domain space minus the dimension of the image space. The diagram commutes. In fact $\chi : F \to F_{pq}$ is the pull back of $\tau_q : F_q \to F_{pq}$ under π. If $(x,y) \in F_{pq}$, then ψ maps the fiber of χ over (x,y) biholomorphically onto the fiber of τ_q over y.

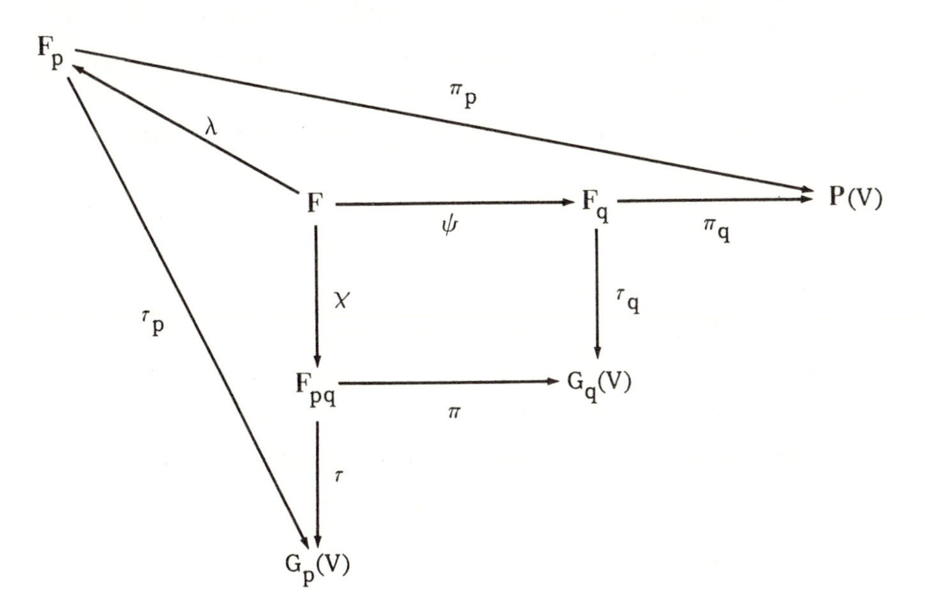

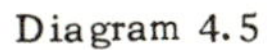
Diagram 4.5

The fiber dimensions of the maps are

τ	π	τ_q	χ	ψ	π_q	π_p	λ	τ_p
(p−q)(p+1)	(n−p)(p−q)	q	q	(n−p)(p−q)	(n−q) q	(n−p) p	(p−q) q	p

The theorems on fiber integration ([25], [26], [33]) imply that

$$\tau_*\pi^*(c_{\mu-q}[q] \wedge c_1[q]^{(p-q)q})$$

$$= \tau_*\pi^*(\tau_{q*}(\pi_q^*(\omega^\mu) \wedge \tau_q^*(c_1[q]^{(p-q)q}))$$

$$= \tau_*\chi_*(\psi^*\pi_q^*(\omega^\mu) \wedge \psi^*\tau_q^*(c_1[q]^{(p-q)q}))$$

$$= \tau_{p*}\lambda_*(\lambda^*\pi_p^*(\omega^\mu) \wedge \psi^*\tau_q^*(c_1[q]^{(p-q)q}))$$

$$= \tau_{p*}(\pi_p^*(\omega^\mu)\lambda_*\psi^*\tau_q^*(c_1[q]^{(p-q)q})) .$$

Here the fiber integral $\alpha = \lambda_*\psi^*\tau_q^*(c_1[q]^{(p-q)q})$ is a function on F_p invariant under the actions of $\mathfrak{U}$. Hence α is constant, which implies that

$$\tau_*\pi^*(c_{\mu-q}[q] \wedge c_1[q]^{(p-q)q}) = \alpha\tau_{p*}(\pi_p^*(\omega^\mu)) = \alpha\, c_{\mu-p}[p] .$$

Therefore only $\alpha = D(q-1, p-1)$ remains to be proved. Take $x \in G_p(V)$. In diagram 4.5 the maps $\tilde{\pi}$ and $\tilde{\tau}$ are projections and j_0 is the inclusion. Define $j_1(y,z) = (x,y,z)$ if $(y,z) \in F_q(E(x))$ and $j_2(z) = (x,z)$ if $z \in P(E(x))$. Then j_0, j_1, j_2 are smooth, injective holomorphic maps. The diagram commutes. Given any $(x,z) \in F_p$, diagram 4.6 can be constructed. Define $N = \lambda^{-1}(x,z)$ and $M = \tilde{\pi}^{-1}(z)$. The restriction $j_1 : M \to N$ is bijective and holomorphic, hence biholomorphic. The hermitian metric on V restricts to the hermitian metric on $E(x)$ which implies that

$$j_0^*(c_1[q,V]) = j_0^*(\omega_{qV}) = \omega_{q,E(x)} = c_1[q,E(x)] .$$

Therefore Proposition 4.6 implies that

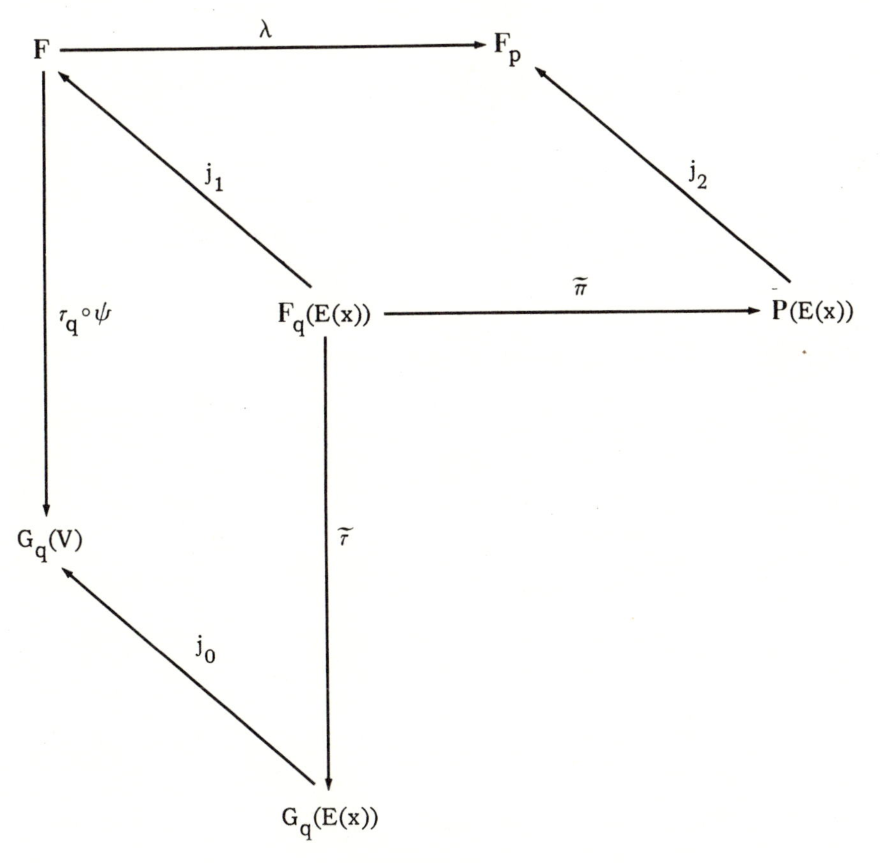

Diagram 4.6

$$\alpha = \int_N (\tau_q \circ \psi)^*(c_1[q,V]^{(p-q)\,q})$$

$$= \int_M j_1^*(\tau_q \circ \psi)^*(c_1[q,V]^{(p-q)\,q})$$

$$= \int_M \tilde{\tau}^* j_0^*(c_1[q,V]^{(p-q)\,q})$$

$$= \int_M \tilde{\tau}^*(c_1[q,E(x)]^{(p-q)\,q})$$

$$= \tilde{\pi}_* \tilde{\tau}^*(c_1[q,E(x)]^{(p-q)\,q})(z) = D(q-1,p-1)\ ;$$

q.e.d.

The power of fiber integration is already evident.

5. THE POINCARÉ DUAL OF A SCHUBERT VARIETY

a) *The volume element of flag spaces*

Let V be a complex vector space of dimension $n+1$. Let $\mathfrak{l}$ be a positive definite hermitian form on V. Take $p \in \mathbb{Z}[0,n]$ and $\mathfrak{a} = (a_0, \cdots, a_p) \in \mathfrak{S}(p,n)$. Let $\pi_\nu : F(\mathfrak{a}) \to G_{\hat{a}_\nu}(V)$ be the projection. Cowen [8] introduced the non-negative form

$$\Omega_{\mathfrak{a}} = \bigwedge_{\nu=0}^{p} \pi_\nu^*(c_{n-\hat{a}_\nu}[\hat{a}_\nu]^{\hat{a}_\nu - \hat{a}_{\nu-1}})$$

of class C^∞ and degree $2d(\mathfrak{a})$ on $F(\mathfrak{a})$. Obviously, $\Omega_{\mathfrak{a}}$ is invariant under the actions of the unitary group.

THEOREM 5.1.
$$\int_{F(\mathfrak{a})} \Omega_{\mathfrak{a}} = 1 .$$

Proof. If $p = 0$, then $F(\mathfrak{a}) = G_{a_0}(V)$ and $\Omega_{\mathfrak{a}} = c_{n-a_0}[a_0]^{a_0+1}$ and Theorem 5.1 holds by Theorem 4.4. Assume $0 < p \leq n$ and suppose that the theorem is proved for $0, 1, \cdots, p-1$. Define $\mathfrak{b} = (a_0, \cdots, a_{p-1}) \in \mathfrak{S}(p-1,n)$ and $s = \hat{a}_p > t = \hat{a}_{p-1}$. All maps in diagram 5.1 are projections and commute with the actions of $GL(V)$. Hence the maps are locally trivial and surjective and have smooth fibers. The diagram commutes. The standard relative product of (π, ρ) is defined by

$$\{((v_0, \cdots, v_{p-1}), (x,y)) \in F(\mathfrak{b}) \times F_{st} \mid \pi(x,y) = \rho(v_0, \cdots, v_{p-1})\}$$
$$= \{((v_0, \cdots, v_{p-1}), (x, v_{p-1})) \in F(\mathfrak{b}) \times F_{s,t} \mid E(v_{p-1}) \subseteq E(x)\}$$

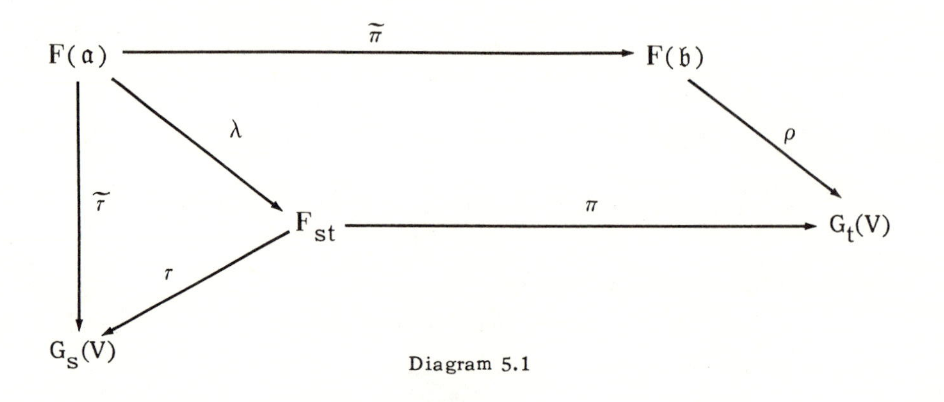

Diagram 5.1

which is biholomorphically equivalent to

$$\{(v_0,\cdots,v_p)\in F(\mathfrak{b})\times G_s(V)\,|\,E(v_{p-1})\subseteq E(v_p)\} = F(\mathfrak{a})\ .$$

Hence $(F(\mathfrak{a}),\tilde{\pi},\lambda)$ is the relative product of (π,ρ). Therefore $\tilde{\pi}$ is the fiber bundle pull back of π by ρ. If $0\le\mu\le p-1$, let $\pi_\mu : F(\mathfrak{a})\to G_{\hat{a}_\mu}(V)$ and $\hat{\pi}_\mu : F(\mathfrak{b})\to G_{\hat{a}_\mu}(V)$ be the projections. Then $\pi_\mu = \hat{\pi}_\mu\circ\tilde{\pi}$. Define $\pi_p = \tilde{\tau}$. If $0\le\mu\le p$, abbreviate $b_\mu = \hat{a}_\mu - \hat{a}_{\mu-1}$. Then $b_p = s-t$. We have

$$\Omega_{\mathfrak{a}} = \left(\bigwedge_{\mu=0}^{p-1}\pi_\mu^*(c_{n-\hat{a}_\mu}[\hat{a}_\mu]^{b_\mu})\right)\wedge\pi_p^*(c_{n-\hat{a}_p}[\hat{a}_p]^{b_p})$$

$$= \left(\bigwedge_{\mu=0}^{p-1}\tilde{\pi}^*(\hat{\pi}_\mu^*(c_{n-\hat{a}_\mu}[\hat{a}_\mu]^{b_\mu})\right)\wedge\tilde{\tau}^*(c_{n-s}[s]^{s-t})$$

$$= \tilde{\pi}^*(\Omega_{\mathfrak{b}})\wedge\lambda^*\tau^*(c_{n-s}[s]^{s-t})\ .$$

Fiber integration, Proposition 4.5 and induction imply that

$$\int_{F(\mathfrak{a})}\Omega_{\mathfrak{a}} = \int_{F(\mathfrak{a})}\tilde{\pi}^*(\Omega_{\mathfrak{b}})\wedge\lambda^*\tau^*(c_{n-s}[s]^{s-t})$$

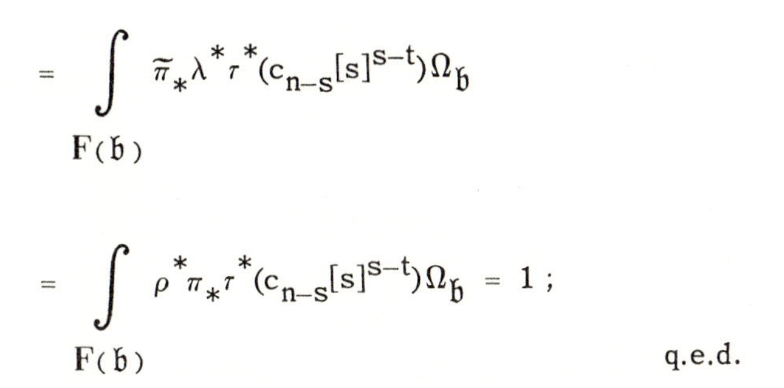

$$= \int_{F(\mathfrak{b})} \tilde{\pi}_* \lambda^* \tau^* (c_{n-s}[s]^{s-t}) \Omega_{\mathfrak{b}}$$

$$= \int_{F(\mathfrak{b})} \rho^* \pi_* \tau^* (c_{n-s}[s]^{s-t}) \Omega_{\mathfrak{b}} = 1 \, ;$$

q.e.d.

Because $\Omega(\mathfrak{a})$ is invariant under the unitary group, $\Omega(\mathfrak{a}) > 0$ at every point of $F(\mathfrak{a})$.

b) *The Chern form of symbol* $\mathfrak{a}$

Take $p \in Z[0,n]$ and $\mathfrak{a} \in \mathfrak{S}(p,n)$. Let $\sigma : S(\mathfrak{a}) \to F(\mathfrak{a})$ and $\pi : S(\mathfrak{a}) \to G_p(V)$ be the projection. Cowen [8] defines the non-negative *Chern form of symbol* $\mathfrak{a}$ by

$$c(\mathfrak{a}) = \pi_* \sigma^* (\Omega_{\mathfrak{a}}) \geq 0 \, . \tag{5.2}$$

Because π has fiber dimension $d(\mathfrak{a}) + \vec{\mathfrak{a}} - d(p,n)$, the form $c(\mathfrak{a})$ has bidegree $(\vec{\mathfrak{a}}^*, \vec{\mathfrak{a}}^*)$ with $\vec{\mathfrak{a}}^* = d(p,n) - \vec{\mathfrak{a}}$. Clearly $c(\mathfrak{a})$ is of class C^∞ and invariant under the action of the unitary group $\mathfrak{U}$, because π, τ commute with this action. Obviously $dc(\mathfrak{a}) = 0$.

THEOREM 5.2. *Take* $v \in F(\mathfrak{a})$. *Let* ψ *be a form of class* C^1 *and bidegree* $(\vec{\mathfrak{a}}, \vec{\mathfrak{a}})$ *on* $G_p(V)$ *with* $d\psi = 0$. *Then*

$$\int_{S(v,\mathfrak{a})} \psi = \int_{G_p(V)} c(\mathfrak{a}) \wedge \psi \, .$$

Therefore, $c(\mathfrak{a})$ *is the Poincaré dual of* $S(v, \mathfrak{a})$ *(Cowen [8]).*

Proof. The function $\sigma_* \pi^* (\psi)$ exists and is of class C^1 on $F(\mathfrak{a})$ with

$d\sigma_*\pi^*(\psi) = \sigma_*\pi^*(d\psi) = 0$. Hence $\sigma_*\pi^*(\psi)$ is a constant γ. Define $N = \sigma^{-1}(v)$ and observe that $S(v, \alpha) = \pi(N)$. Hence

$$\gamma = \sigma_*\pi^*(\psi)(v) = \int_N \pi^*(\psi) = \int_{S(v,\alpha)} \psi .$$

Fiber integration (Tung [33]) implies that

$$\int_{G_p(V)} c(\alpha) \wedge \psi = \int_{G_p(V)} \pi_*\sigma^*(\Omega_\alpha) \wedge \psi = \int_{S(\alpha)} \sigma^*(\Omega_\alpha) \wedge \pi^*(\psi)$$

$$= \int_{F(\alpha)} \sigma_*\pi^*(\psi)\Omega_\alpha = \gamma ;$$

q.e.d.

LEMMA 5.3 (Crofton's formula). *Take* $q \in Z[0,n]$ *and* $v \in G_q(V)$. *Let* ψ *be a form of class* C^1 *and bidegree* (q,q) *on* $P(V)$ *with* $d\chi = 0$. *Then*

$$\int_{P(V)} \omega^{n-q} \wedge \chi = \int_{\ddot{E}(v)} \chi .$$

Proof. In Theorem 5.2 take $p = 0$ and $\alpha = (q)$. Take $v \in F(\alpha) = G_q(V)$. Then $S(v, \alpha) = \ddot{E}(v)$. Because $c(q)$ is invariant under the action of the unitary group, and ω^{n-q} generates $H^{2n-2q}(P(V), C)$, a constant γ exists such that $c(q) = \gamma\omega^{n-q}$. Theorem 5.2 implies that

$$\int_{\ddot{E}(v)} \chi = \gamma \int_{P(V)} \omega^{n-q} \wedge \chi .$$

Here γ is independent of χ. The choice $\chi = \omega^q$ shows $\gamma = 1$; q.e.d.

Of course there are many proofs of Lemma 5.3. For instance the un-integrated First Main Theorem [25] Theorems 4.5 and 4.6 imply Lemma 5.3. A direct proof is based on cohomology and Stokes' Theorem. Because ω^q generates $H^{2q}(P(V), \mathbb{C})$, there exists a constant a and a form ξ such that $\chi = a\omega^q + d\xi$. Stokes' Theorem implies that

$$\int_{\ddot{E}(v)} \chi = a \int_{\ddot{E}(v)} \omega^q = a = \int_{P(V)} a\omega^n = \int_{P(V)} \omega^{n-q} \wedge a\omega^q$$

$$= \int_{P(V)} \omega^{n-q} \wedge \chi .$$

PROPOSITION 5.4. *If* $p \in Z[0,n]$ *and* $h \in Z[0,n-p]$, *then*

$$c_h[p] = c(n-p-h, n-p, \cdots, n-p) .$$

Proof. Define $s = n-p-h$ and $\mathfrak{a} = (s, n-p, \cdots, n-p)$. Take $v = (v_0, \cdots, v_p) \in F(\mathfrak{a})$. Let $\pi: F_p \to P(V)$ and $\tau: F_p \to G_p(V)$ be the projections. By (2.4), we have $S(v, \mathfrak{a}) = \tau\pi^{-1}(v_0)$. Observe that $\vec{\mathfrak{a}} = (n-p)(p+1) - h$. Let ψ be a form of class C^1 and bidegree $(\vec{\mathfrak{a}}, \vec{\mathfrak{a}})$ on $G_p(V)$ with $d\psi = 0$. Then $\pi_*\tau^*(\psi)$ is a form of class C^1 and bidegree (s,s) on $P(V)$ with $d\pi_*\tau^*(\psi) = \pi_*\tau^*(d\psi) = 0$. Hence

$$\int_{\ddot{E}(v_0)} \pi_*\tau^*(\psi) = \int_{P(V)} \omega^{p+h} \wedge \pi_*\tau^*(\psi) = \int_{F_p} \pi^*(\omega^{p+h}) \wedge \tau^*(\psi)$$

$$= \int_{G_p(V)} \tau_*\pi^*(\omega^{p+h}) \wedge \psi = \int_{G_p(V)} c_h[p] \wedge \psi .$$

Let X and Y be complex spaces. A proper, surjective holomorphic map $f: X \to Y$ is called a *modification*, if there exists a thin analytic subset A of Y such that $B = f^{-1}(A)$ is thin in X and such that $f:(X-B) \to (Y-A)$ is biholomorphic.

Because $\ddot{E}(v_0)$ is a compact, connected, complex manifold of dimension s and because the fibers of π are connected complex manifolds of dimension $(n-p)p$, the analytic set

$$N = \pi^{-1}(\ddot{E}(v_0)) = \{(x,y) \in G_p(V) \times \ddot{E}(v_0) \mid y \in \ddot{E}(x)\}$$

is irreducible and has dimension $(n-p)p+s = \vec{a}$ by Lemma 1.2. Example 6 of §2 implies that

$$\tau(N) = \{x \in G_p(V) \mid \ddot{E}(x) \cap \ddot{E}(v_0) \neq \emptyset\} = S(v, a) \ .$$

The restriction $\hat{\tau} = \tau : N \to S(v, a)$ is holomorphic and surjective. Each fiber $\hat{\tau}^{-1}(x) = \ddot{E}(x) \cap \ddot{E}(v_0)$ is a connected manifold. Hence, because N and $S(v, a)$ have the same dimension $\vec{a}$ and are irreducible, $\hat{\tau}$ is a modification. In the commutative diagram 5.3, the maps j, $\tilde{j}$, $\hat{j}$ are inclusions and τ, $\hat{\tau}$, π, $\tilde{\pi}$ are projections.

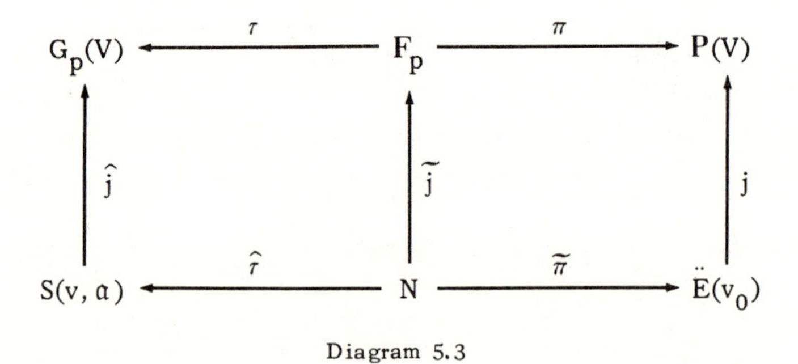

Diagram 5.3

Here $\tilde{\pi}: N \to \ddot{E}(v_0)$ is the pull back of the holomorphic fiber bundle $\pi : F_p \to P(V)$ under the inclusion j . Therefore

$$j^*\pi_*\tau^*(\psi) = \tilde{\pi}_*\tilde{j}^*\tau^*(\psi) = \tilde{\pi}_*\hat{\tau}^*\hat{j}^*(\psi) .$$

Since $\hat{\tau}$ is a modification, this implies that

$$\int_{\ddot{E}(v_0)} \pi_*\tau^*(\psi) = \int_{\ddot{E}(v_0)} j^*\pi_*\tau^*(\psi) = \int_{\ddot{E}(v_0)} \tilde{\pi}_\nu\hat{\tau}^*\hat{j}^*(\psi)$$

$$= \int_N \hat{\tau}^*\hat{j}^*(\psi) = \int_{S(v,\alpha)} \psi .$$

Therefore

$$\int_{G_p(V)} c_h[p] \wedge \psi = \int_{S(v,\alpha)} \psi$$

for all forms ψ of class C^1 and bidegree $(\vec{\alpha}, \vec{\alpha})$ with $d\psi = 0$. Because the cohomology class of the Poincaré dual is unique, and because $c(\alpha)$ and $c_h[p]$ are invariant under $\mathfrak{U}(V)$, we have $c(\alpha) = c_h[p]$; q.e.d.

6. MATSUSHIMA'S THEOREM

a) *Double intersection*

Let V be a complex vector space of dimension $n+1 \geq 1$. Let $\mathfrak{l}$ be a positive definite hermitian form on V. Take $p \in \mathbf{Z}[0,n]$. Take $\mathfrak{a}$ and $\mathfrak{b}$ in $\mathfrak{S}(p,n)$. Then

$$S(\mathfrak{a},\mathfrak{b}) = \{(x,v,w) \mid (x,v) \in \mathbf{F}(\mathfrak{a}) \text{ and } (x,w) \in \mathbf{F}(\mathfrak{b})\}$$

is an analytic subset of $G_p(V) \times \mathbf{F}(\mathfrak{a}) \times \mathbf{F}(\mathfrak{b})$. All maps in the *double intersection diagram* 6.1 are projections.

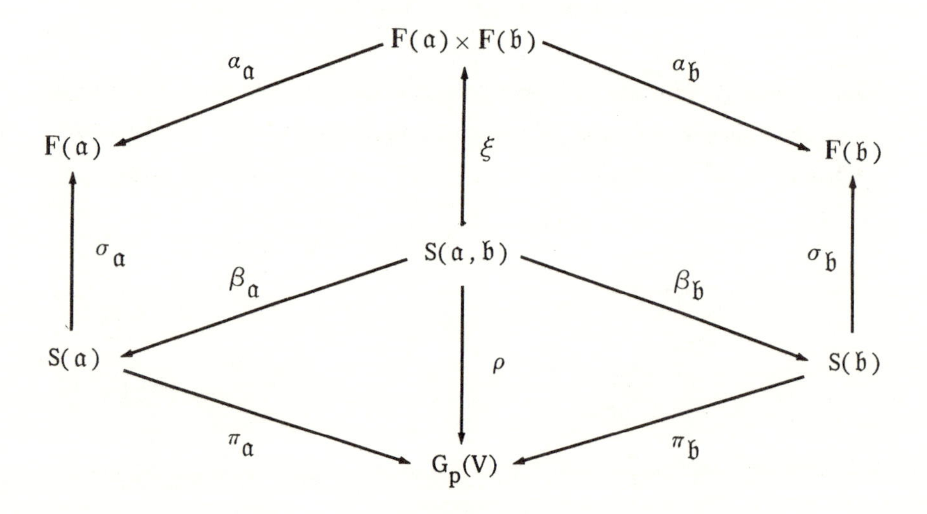

Diagram 6.1

The diagram commutes. $(S(\mathfrak{a},\mathfrak{b}), \beta_{\mathfrak{a}}, \beta_{\mathfrak{b}})$ is the relative product of $(\pi_{\mathfrak{b}}, \pi_{\mathfrak{a}})$. If $(v,w) \in \mathbf{F}(\mathfrak{a}) \times \mathbf{F}(\mathfrak{b})$, then

$$\xi^{-1}(v,w) = (S(v,\mathfrak{a}) \cap S(w,\mathfrak{a})) \times \{(v,w)\} . \tag{6.2}$$

Hence the restriction

$$\rho : \xi^{-1}(v,w) \to S(v,\mathfrak{a}) \cap S(w,\mathfrak{a}) \tag{6.3}$$

is biholomorphic.

LEMMA 6.1. *The analytic set* $S(\mathfrak{a},\mathfrak{b})$ *is irreducible with*

$$d(\mathfrak{a},\mathfrak{b}) = \dim S(\mathfrak{a},\mathfrak{b}) = d(\mathfrak{a}) + d(\mathfrak{b}) + \vec{\mathfrak{a}} + \vec{\mathfrak{b}} - d(p,n) .$$

Also $\beta_{\mathfrak{a}}$ *and* $\beta_{\mathfrak{b}}$ *are surjective and locally trivial, and have irreducible fibers.*

Proof. Because $(S(\mathfrak{a},\mathfrak{b}),\beta_{\mathfrak{a}},\beta_{\mathfrak{b}})$ is the relative product of $(\pi_{\mathfrak{b}},\pi_{\mathfrak{a}})$, the maps $\beta_{\mathfrak{a}}$ and $\beta_{\mathfrak{b}}$ are locally trivial and surjective. The fibers of $\beta_{\mathfrak{a}}$ (respectively $\beta_{\mathfrak{b}}$) are isomorphic to the fibers of $\pi_{\mathfrak{b}}$ (respectively $\pi_{\mathfrak{a}}$). Hence the fibers of $\beta_{\mathfrak{a}}$ and $\beta_{\mathfrak{b}}$ are irreducible. Since $S(\mathfrak{a})$ and $S(\mathfrak{b})$ are irreducible; also $S(\mathfrak{a},\mathfrak{b})$ is irreducible by Lemma 1.2 with

$$\begin{aligned}
\dim S(\mathfrak{a},\mathfrak{b}) &= \dim S(\mathfrak{a}) + \text{fib.}\dim \beta_{\mathfrak{a}} \\
&= \dim S(\mathfrak{a}) + \text{fib.}\dim \pi_{\mathfrak{b}} \\
&= d(\mathfrak{a}) + \vec{\mathfrak{a}} + d(\mathfrak{b}) + \vec{\mathfrak{b}} - d(p,n) ;
\end{aligned}$$

q.e.d.

Observe that $\dim F(\mathfrak{a}) \times F(\mathfrak{b}) = d(\mathfrak{a}) + d(\mathfrak{b})$. Hence

$$\dim S(\mathfrak{a},\mathfrak{b}) - \dim F(\mathfrak{a}) \times F(\mathfrak{b}) = \vec{\mathfrak{a}} + \vec{\mathfrak{b}} - d(p,n) . \tag{6.4}$$

PROPOSITION 6.2. $c(\mathfrak{a}) \wedge c(\mathfrak{b}) = \rho_* \xi^*(\alpha_{\mathfrak{a}}^*(\Omega_{\mathfrak{a}}) \cap \alpha_{\mathfrak{b}}^*(\Omega_{\mathfrak{b}}))$.

Proof. Theorem 4.1 implies that

$$\begin{aligned}
c(a) \wedge c(b) &= \pi_{\mathfrak{a}*}\sigma_{\mathfrak{a}}^*(\Omega_{\mathfrak{a}}) \wedge \pi_{\mathfrak{b}*}\sigma_{\mathfrak{b}}^*(\Omega_{\mathfrak{b}}) \\
&= \rho_*(\beta_{\mathfrak{a}}^*\sigma_{\mathfrak{a}}^*(\Omega_{\mathfrak{a}}) \wedge \beta_{\mathfrak{b}}^*\sigma_{\mathfrak{b}}^*(\Omega_{\mathfrak{b}})) \\
&= \rho_*\xi^*(\alpha_{\mathfrak{a}}^*(\Omega_{\mathfrak{a}}) \wedge \alpha_{\mathfrak{b}}^*(\Omega_{\mathfrak{b}})) ;
\end{aligned}$$

q.e.d.

LEMMA 6.3. *If* ξ *is surjective, then* $\mathfrak{a} \geq \mathfrak{b}^*$.

Proof. Define $\mathfrak{c} = \mathfrak{b}^*$. Then $c_q = n-p-b_{p-q}$ for $q = 0,\cdots,p$. Let $e_0,\cdots,e_n$ be a base of V . Define

$$v_q = P(e_0 \wedge \cdots \wedge e_{\hat{a}_q}) \qquad w_q = P(e_n \wedge \cdots \wedge e_{n-\hat{b}_q})$$

for $q = 0,\cdots,p$. Then $v = (v_0,\cdots,v_p) \in F(\mathfrak{a})$ and $w = (w_0,\cdots,w_p) \in F(\mathfrak{b})$. Also $n - \hat{b}_{p-q} = c_q + q = \hat{c}_q$ with

$$E(v_q) = \mathbb{C}\, e_0 + \cdots + \mathbb{C}\, e_{\hat{a}_q}$$

$$E(w_{p-q}) = \mathbb{C}\, e_n + \cdots + \mathbb{C}\, e_{\hat{c}_q} .$$

Because ξ is surjective, $x \in G_p(V)$ exists such that $(x, v, w) \in S(\mathfrak{a}, \mathfrak{b})$. Hence

$$\dim E(x) \cap E(v_q) \geq q+1 \qquad \dim E(x) \cap E(w_{p-q}) \geq p-q+1 .$$

In particular $E(x) \subseteq E(v_p)$ and $E(x) \subseteq E(w_p)$. Therefore

$$\dim E(x) \cap E(v_q) \cap E(w_{p-q}) \geq q+1 + p-q + 1 - p-1 = 1 .$$

Hence $\dim E(v_q) \cap E(w_{p-q}) \geq 1$ and this implies that $\hat{c}_q \leq \hat{a}_q$ or $c_q \leq a_q$ for $q = 0,\cdots,p$. Therefore $\mathfrak{b}^* = \mathfrak{c} \leq \mathfrak{a}$; q.e.d.

PROPOSITION 6.4. *Take* $\mathfrak{a}$ *and* $\mathfrak{b}$ *in* $\mathfrak{S}(p,n)$. *Assume that* $c(\mathfrak{a}) \wedge c(\mathfrak{b}) \neq 0$. *Then* $\mathfrak{a} \geq \mathfrak{b}^*$ *and* $\vec{\mathfrak{a}} + \vec{\mathfrak{b}} \geq d(p,n)$.

Proof. Assume that ξ is not surjective. Then $T = \xi(S(\mathfrak{a}, \mathfrak{b}))$ is a thin analytic subset of $F(\mathfrak{a}) \times F(\mathfrak{b})$ and ξ restricts to $\xi_0 : S(\mathfrak{a}, \mathfrak{b}) \to T$. Let $\iota : S(\mathfrak{a}, \mathfrak{b}) \to F(\mathfrak{a}) \times F(\mathfrak{b})$ be the inclusion map. Then $\xi = \iota \circ \xi_0$. Because $\alpha_{\mathfrak{a}}^*(\Omega_{\mathfrak{a}}) \wedge \alpha_{\mathfrak{b}}^*(\Omega_{\mathfrak{b}})$ has degree $2(d(\mathfrak{a})+d(\mathfrak{b})) > \dim T$, we have $\iota^*(\alpha_{\mathfrak{a}}^*(\Omega_{\mathfrak{a}}) \wedge \alpha_{\mathfrak{b}}^*(\Omega_{\mathfrak{b}}) = 0$. Hence

$$\xi(\alpha_{\mathfrak{a}}^{*}(\Omega_{\mathfrak{a}}) \wedge \alpha_{\mathfrak{b}}^{*}(\Omega_{\mathfrak{b}})) = \xi_0^{*} \iota^{*}(\alpha_{\mathfrak{a}}^{*}(\Omega_{\mathfrak{a}}) \wedge \alpha_{\mathfrak{b}}^{*}(\Omega_{\mathfrak{b}})) = 0 .$$

By Proposition 6.2 $c(\mathfrak{a}) \wedge c(\mathfrak{b}) = 0$ contrary to the assumption. Therefore ξ is surjective, and consequently $\mathfrak{a} \geq \mathfrak{b}^{*}$ by Lemma 6.3. Hence $\vec{\mathfrak{a}} \geq \vec{\mathfrak{b}}^{*} = d(p,n) - \vec{\mathfrak{b}}$; q.e.d.

The case of an injective map ξ needs more preparations.

b) *General position*

Take $p \in \mathbf{Z}[0,n]$ and $\mathfrak{a} \in \mathfrak{S}(p,n)$. Then $(x,v) \in S(\mathfrak{a})$ is said to be in *general position* if $\dim E(x) \cap E(v_q) = q+1$ for all $q = 0, 1, \cdots, p$. Let $H(\mathfrak{a})$ be the set of all $(x,v) \in S(\mathfrak{a})$ in general position.

LEMMA 6.5. *$H(\mathfrak{a})$ is a non-empty Zariski open subset of $S(\mathfrak{a})$. In particular, $H(\mathfrak{a})$ is dense in $S(\mathfrak{a})$. If $v \in F(\mathfrak{a})$, then $H(\mathfrak{a}) \cap \pi^{-1}(v)$ is dense in $\pi^{-1}(v)$.*

Proof. By Lemma 1.5 the set L_q of all $(x,v) \in S(\mathfrak{a})$ such that $\dim E(x) \cap E(v_q) \geq q+2$ is analytic. Hence $L = L_0 \cup \cdots \cup L_p$ is analytic with $H(\mathfrak{a}) = S(\mathfrak{a}) - L$. Hence $H(\mathfrak{a})$ is Zariski open. Take $v \in F(\mathfrak{a})$. A complete flag $w \in F_n$ exists such that $v = \phi_{\mathfrak{a}}(w)$ where $\phi_{\mathfrak{a}} : F_n \to F(\mathfrak{a})$ is the projection. Then $S(v, \mathfrak{a}) = S\langle w, \mathfrak{a}\rangle$. By Lemma 2.2, $S^{*}\langle w, \mathfrak{a}\rangle \subseteq H(\mathfrak{a})$. Hence $H(\mathfrak{a}) \cap \pi^{-1}(v) \neq \emptyset$ is Zariski open and dense in the irreducible analytic set $\pi^{-1}(v)$. In particular $H(\mathfrak{a}) \neq \emptyset$. Hence $H(\mathfrak{a})$ is dense in the irreducible analytic set $S(\mathfrak{a})$; q.e.d.

Take $(x,v) \in S(a)$. Then $(\mathfrak{x}, \mathfrak{v})$ is called a *representation* of (x,v) if the following conditions are satisfied:

(1) We have $\mathfrak{x} = (\mathfrak{x}_0, \cdots, \mathfrak{x}_p)$, where $\mathfrak{x}_0, \cdots, \mathfrak{x}_p$ are linearly independent vectors in V with $x = \mathbf{P}(\mathfrak{x}_0 \wedge \cdots \wedge \mathfrak{x}_p)$.

(2) We have $\mathfrak{v} = (\mathfrak{v}_0, \cdots, \mathfrak{v}_{\hat{a}_p})$, where $\mathfrak{v}_0, \cdots, \mathfrak{v}_{\hat{a}_p}$ are linear by independent vectors in V, with $v_q = \mathbf{P}(\mathfrak{v}_0 \wedge \cdots \wedge \mathfrak{v}_{\hat{a}_q})$ for $q = 0, \cdots, p$. Moreover $v = (v_0, \cdots, v_p)$.

(3) We have $\mathfrak{x}_q = \mathfrak{v}_{\hat{a}_q}$ for $q = 0, \cdots, p$.

LEMMA 6.5. *Take* $(x,v) \in S(\mathfrak{a})$. *Then* (x,v) *is in general position if and only if a representation* $(\mathfrak{x}, \mathfrak{v})$ *of* (x,v) *exists.*

Proof. a) *Assume that* (x,v) *is in general position.* A linear subspace L_q of V of dimension $\hat{a}_q - \hat{a}_{q-1}$ exists such that $E(v_q) = E(v_{q-1}) \oplus L_q$ for $q = 0, 1, \cdots, p$. Because $\dim E(x) \cap E(v_{q-1}) = q$ and because $\dim E(x) \cap E(v_q) = q+1$, we have $\dim E(x) \cap L_q = 1$. Hence linearly independent vectors $\mathfrak{v}_0, \cdots, \mathfrak{v}_{\hat{a}_p}$ can be taken such that $\mathfrak{v}_0, \cdots, \mathfrak{v}_{\hat{a}_q}$ span $E(v_q)$ and $\mathfrak{v}_{\hat{a}_q}$ spans $E(x) \cap E(v_q)$ for each $q = 0, \cdots, p$. Define $\mathfrak{v} = (\mathfrak{v}_0, \cdots, \mathfrak{v}_{\hat{a}_p})$ and $\mathfrak{x}_q = \mathfrak{v}_{\hat{a}_q}$. Then $\mathfrak{x}_0, \cdots, \mathfrak{x}_p$ are linearly independent and span $E(x)$. Hence $x = \mathbf{P}(\mathfrak{x}_0 \wedge \cdots \wedge \mathfrak{x}_p)$. Also $v_q = \mathbf{P}(\mathfrak{v}_0 \wedge \cdots \wedge \mathfrak{v}_{\hat{a}_q})$ for $q = 0, \cdots, p$. Define $\mathfrak{x} = (\mathfrak{x}_0, \cdots, \mathfrak{x}_p)$. Then $(\mathfrak{x}, \mathfrak{v})$ is a representation of (x,v).

b) *Assume that* $(\mathfrak{x}, \mathfrak{v})$ *is a representation of* (x,v). Take $q \in \mathbf{Z}[0,p]$. If $q = p$, then $E(x) \subseteq E(v_p)$ and $\dim E(x) \cap E(v_p) = p+1$. Assume that $q < p$. Let M_q be the linear subspace spanned by $\mathfrak{v}_\mu$ for $\mu = \hat{a}_q + 1, \hat{a}_q + 2, \cdots, \hat{a}_p$. Then $\mathfrak{x}_\mu \in E(v_q)$ for $\mu = 0, \cdots, q$ and $\mathfrak{x}_\mu \in M_q$ for $\mu = q+1, \cdots, p$ with $M_q \cap E(v_q) = \emptyset$. Hence

$$E(x) = (E(x) \cap E(v_q)) \oplus (E(x) \cap M_q) .$$

Because $\dim E(x) \cap E(v_q) \geq q+1$ and $\dim E(x) \cap M_q \geq p-q$ while $\dim E(x) = p+1$, we obtain $\dim E(x) \cap E(v_q) = q+1$. Therefore (x,v) is in general position; q.e.d.

Take $\mathfrak{a}$ and $\mathfrak{b}$ in $\mathfrak{S}(p,n)$. Then $(x,v,w) \in S(\mathfrak{a}, \mathfrak{b})$ is said to be in *general position* if the following conditions are satisfied:

(G1) $(x,v) \in S(\mathfrak{a})$ and $(x,w) \in S(\mathfrak{a})$ are in general position.

(G2) The vector space $L_q = E(x) \cap E(v_q) \cap E(w_{p-q})$ has dimension 1 for $q = 0, \cdots, p$.

(G3) If $0 \leq m < q < p$, then $L_m \cap L_q = 0$.
Obviously (G3) is equivalent to

(G3′) $E(x) = L_0 \oplus \cdots \oplus L_p$.

Let $H(\mathfrak{a}, \mathfrak{b})$ be the set of all $(x,v,w) \in S(\mathfrak{a}, \mathfrak{b})$ in general position.

Take $(x,v,w) \in S(\mathfrak{a}, \mathfrak{b})$. Then $(\mathfrak{x}, \mathfrak{v}, \mathfrak{w})$ is said to be a *representation* of (x,v,w) if the following conditions are satisfied:

(R1) $(\mathfrak{x}, \mathfrak{v})$ is a representation of (x,v) with $\mathfrak{x} = (\mathfrak{x}_0, \cdots, \mathfrak{x}_p)$.

(R2) For $q = 0, \cdots, p$ define $\mathfrak{y}_q = \mathfrak{x}_{p-q}$. Define $\mathfrak{y} = (\mathfrak{y}_0, \cdots, \mathfrak{y}_p)$. Then $(\mathfrak{y}, \mathfrak{w})$ is a representation of (x,w).

LEMMA 6.6. *Take* $(x,v,w) \in S(\mathfrak{a}, \mathfrak{b})$. *Then* (x,v,w) *is in general position if and only if a representation* $(\mathfrak{x}, \mathfrak{v}, \mathfrak{w})$ *of* (x,v,w) *exists.*

Proof. a). *Assume that* (x,v,w) *is in general position.* The vector space L_q defined in (G2) has dimension 1. Take $0 \neq \mathfrak{x}_q \in L_q$. By (G3′), the vectors $\mathfrak{x}_0, \cdots, \mathfrak{x}_p$ are linearly independent with $x = \mathbf{P}(\mathfrak{x}_0 \wedge \cdots \wedge \mathfrak{x}_p)$. Define $\mathfrak{x} = (\mathfrak{x}_0, \cdots, \mathfrak{x}_p)$. Now, linearly independent vectors $\mathfrak{v}_0, \cdots, \mathfrak{v}_{\hat{a}_q}$ shall be constructed such that $\mathfrak{v}_0, \cdots, \mathfrak{v}_{\hat{a}_\mu}$ span $E(v_\mu)$ with $\mathfrak{v}_{\hat{a}_\mu} = \mathfrak{x}_\mu$ for $\mu = 0, \cdots, q$. If $q = 0$, this is trivial since $\mathfrak{x}_0 \in E(v_0)$. Assume that $\mathfrak{v}_0, \cdots, \mathfrak{v}_{\hat{a}_{q-1}}$ are constructed with $q \leq p$. Assume that $\mathfrak{x}_q \in E(v_{q-1})$. Because $\mathfrak{x}_q \in E(x) \cap E(w_{p-q})$ we have $0 \neq \mathfrak{x}_q \in E(x) \cap E(v_{q-1}) \cap E(w_{p-q}) \subseteq L_q$. Since $\dim L_q = 1$, we have

$$L_q = E(x) \cap E(v_{q-1}) \cap E(w_{p-q}) \subseteq E(x) \cap E(v_{q-1}) \cap E(w_{p-q+1}) = L_{q-1}$$

which is impossible. Therefore $\mathfrak{x}_q \in E(v_q) - E(v_{q-1})$. Vectors $\mathfrak{v}_\mu$ exist for $\mu = \hat{a}_{q-1} + 1, \cdots, \hat{a}_q$ such that $\mathfrak{x}_q = \mathfrak{v}_{\hat{a}_q}$ and such that $\mathfrak{v}_0, \cdots, \mathfrak{v}_{\hat{a}_q}$ span $E(v_q)$. By induction linearly independent vectors $\mathfrak{v}_0, \cdots, \mathfrak{v}_{\hat{a}_p}$ are constructed such that $\mathfrak{x}_q = \mathfrak{v}_{\hat{a}_q}$ and $v_q = E(\mathfrak{v}_0 \wedge \cdots \wedge \mathfrak{v}_{\hat{a}_q})$ for $q = 0, \cdots, p$. Define $\mathfrak{v} = (\mathfrak{v}_0, \cdots, \mathfrak{v}_{\hat{a}_p})$. Then $(\mathfrak{x}, \mathfrak{v})$ is a representation of (x,v).

Define $\mathfrak{y}_q = \mathfrak{x}_{p-q}$ for $q = 0,\cdots,p$. Define $\mathfrak{y} = (\mathfrak{y}_0,\cdots,\mathfrak{y}_p)$. Then $x = P(\mathfrak{y}_0 \wedge \cdots \wedge \mathfrak{y}_q)$ and

$$0 \neq \mathfrak{y}_q \in E(x) \cap E(w_q) \cap E(v_{p-q}) .$$

Hence the same reasoning as above produces linearly independent vectors $\mathfrak{w}_0,\cdots,\mathfrak{w}_{\hat{a}_p}$ such that $\mathfrak{w} = (\mathfrak{w}_0,\cdots,\mathfrak{w}_{\hat{a}_p})$ and such that $(\mathfrak{y},\mathfrak{w})$ is a representation of (x,w). Then $(\mathfrak{x},\mathfrak{v},\mathfrak{w})$ is a representation of (x,v,w).

b) *Let* $(\mathfrak{x},\mathfrak{v},\mathfrak{w})$ *be a representation of* (x,v,w). Then $\mathfrak{x} = (\mathfrak{x}_0,\cdots,\mathfrak{x}_p)$. Define $\mathfrak{y}_q = \mathfrak{x}_{p-q}$ for $q = 0,\cdots,p$. Define $\mathfrak{y} = (\mathfrak{y}_0,\cdots,\mathfrak{y}_p)$. Then $(\mathfrak{x},\mathfrak{v})$ and $(\mathfrak{y},\mathfrak{w})$ are representations of (x,v) and (x,w) respectively. Therefore (x,v) and (x,w) are in general position. We have $\mathfrak{x}_\mu \in E(x) \cap E(v_q)$ for $\mu = 0,\cdots,q$ with $\dim E(x) \cap E(v_q) = q+1$. The vectors $\mathfrak{x}_0,\cdots,\mathfrak{x}_q$ are linearly independent. Therefore

$$E(x) \cap E(v_q) = \mathbf{C}\,\mathfrak{x}_0 + \cdots + \mathbf{C}\,\mathfrak{x}_q .$$

By symmetry $E(x) \cap E(w_q) = \mathbf{C}\,\mathfrak{y}_0 + \cdots + \mathbf{C}\,\mathfrak{y}_q$. Therefore

$$E(x) \cap E(w_q) = \mathbf{C}\,\mathfrak{x}_p + \cdots + \mathbf{C}\,\mathfrak{x}_q$$

$$L_q = E(x) \cap E(v_q) \cap E(w_{p-q}) = \mathbf{C}\,\mathfrak{x}_q .$$

Hence $\dim L_q = 1$ for $q = 0,\cdots,p$ and $L_m \cap L_q = 0$ if $0 \leq m < q \leq p$. Therefore (x,v,w) is in general position.

LEMMA 6.7. *Take* $\mathfrak{a}$ *and* $\mathfrak{b}$ *in* $\mathfrak{S}(p,n)$. *Then* $H(\mathfrak{a},\mathfrak{b})$ *is a non-empty Zariski open subset of* $S(\mathfrak{a},\mathfrak{b})$. *In particular* $H(\mathfrak{a},\mathfrak{b})$ *is dense in* $S(\mathfrak{a},\mathfrak{b})$.

Proof. Consider diagram 6.1. Since $N_{\mathfrak{a}} = S(\mathfrak{a}) - H(\mathfrak{a})$ and $N_{\mathfrak{b}} = S(\mathfrak{b}) - H(\mathfrak{b})$ are thin analytic subsets of $S(\mathfrak{a})$ and $S(\mathfrak{b})$ respectively

$$N_0 = \beta_{\mathfrak{a}}^{-1}(N_{\mathfrak{a}}) \cup \beta_{\mathfrak{b}}^{-1}(N_{\mathfrak{b}})$$

is a thin analytic subset of $S(\mathfrak{a},\mathfrak{b})$. Define $H_0 = S(\mathfrak{a},\mathfrak{b}) - N_0$. For $(x,v,w) \in S(\mathfrak{a},\mathfrak{b})$ define

$$L_q(x,v,w) = E(x) \cap E(v_q) \cap E(w_{p-q}) .$$

By Lemma 1.5, the sets

$$N_1 = \bigcup_{q=0}^{p} \{(x,v,w) \mid \dim L_q(x,v,w) \geq 2\}$$

$$N_2 = \bigcup_{m \neq q} \{(x,v,w) \mid \dim L_m(x,v,w) \cap L_q(x,v,w) \geq 1\}$$

are analytic subsets of $S(\mathfrak{a},\mathfrak{b})$. Hence $N = N_0 \cup N_1 \cup N_2$ is analytic. Then $H(\mathfrak{a},\mathfrak{b}) = S(\mathfrak{a},\mathfrak{b}) - N$. Only $H(\mathfrak{a},\mathfrak{b}) \neq \emptyset$ remains to be shown.

Take $(x,v,w) \in S(\mathfrak{a},\mathfrak{b})$. Then $(\mathfrak{x},\mathfrak{v})$ and $(\mathfrak{y},\mathfrak{w})$ is called a *double representation* of order j if and only if $(\mathfrak{x},\mathfrak{v})$ and $(\mathfrak{y},\mathfrak{w})$ are representations of (x,v) and (x,w) respectively, and if $\mathfrak{x} = (\mathfrak{x}_0,\cdots,\mathfrak{x}_p)$ and $\mathfrak{y} = (\mathfrak{y}_0,\cdots,\mathfrak{y}_p)$ with $\mathfrak{x}_q = \mathfrak{y}_{p-q}$ for $q = 0,1,\cdots,j$. Let $H(j)$ be the set of all $(x,v,w) \in S(\mathfrak{a},\mathfrak{b})$ for which a double representation of order j exists. Clearly

$$H_0 = H(-1) \supseteq H(0) \supseteq H(1) \supseteq \cdots \supseteq H(p) = H(\mathfrak{a},\mathfrak{b}) .$$

Here $H(-1) = H_0 \neq \emptyset$. Assume that $0 \leq j \leq p$ and $H(j-1) \neq \emptyset$. Then $H(j) \neq \emptyset$ shall be proved. Take $(x,v,w) \in H(j-1)$. Let $(\mathfrak{x},\mathfrak{v})$ and $(\mathfrak{y},\mathfrak{w})$ be a double representation of order $j-1$ of (x,v,w). Then $\mathfrak{x} = (\mathfrak{x}_0,\cdots,\mathfrak{x}_p)$ and $\mathfrak{y} = (\mathfrak{y}_0,\cdots,\mathfrak{y}_p)$ with $\mathfrak{x}_q = \mathfrak{y}_{q-p}$ for $q = 0,1,\cdots,j-1$. Also $\mathfrak{v} = (\mathfrak{v}_0,\cdots,\mathfrak{v}_{\hat{a}_p})$ and $\mathfrak{w} = (\mathfrak{w}_0,\cdots,\mathfrak{w}_{\hat{b}_p})$ with $\mathfrak{x}_q = \mathfrak{v}_{\hat{a}_q}$ and $\mathfrak{y}_q = \mathfrak{w}_{\hat{b}_q}$ for $q = 0,1,\cdots,p$. Because $\mathfrak{x}_0,\cdots,\mathfrak{x}_p$ and $\mathfrak{y}_0,\cdots,\mathfrak{y}_p$ each span $E(x)$, numbers $h_q \in \mathbb{C}$ exist such that $\mathfrak{x}_j = h_0\mathfrak{y}_0 + \cdots + h_p\mathfrak{y}_p$.

CLAIM 1. W.l.o.g. $h_q = 0$ for $q = p-j+1,\cdots,p$ can be assumed.

Proof of Claim 1. If $j = 0$, claim 1 is vacuous. Assume that $j > 0$.

Define $\mathfrak{x}'_q = \mathfrak{x}_q$ for $q \neq j$ and

$$\mathfrak{x}'_j = \mathfrak{x}_j - \sum_{q=0}^{j-1} h_{p-q}\mathfrak{y}_{p-q} = \mathfrak{x}_j - \sum_{q=0}^{j-1} h_{p-q}\mathfrak{x}_q .$$

Then $\mathfrak{x}_0 \wedge \cdots \wedge \mathfrak{x}_p = \mathfrak{x}'_0 \wedge \cdots \wedge \mathfrak{x}'_p$. Define $\mathfrak{v}'_\mu = \mathfrak{v}_\mu$ if $\mu \neq \hat{a}_j$ and $\mathfrak{v}'_{\hat{a}_j} = \mathfrak{x}'_j$. Because $\mathfrak{x}_q = \mathfrak{v}_{\hat{a}_q} = \mathfrak{v}'_{\hat{a}_q}$ for $q = 0,\cdots,j-1$, this implies that $v_q = P(\mathfrak{v}'_0 \wedge \cdots \wedge \mathfrak{v}'_{\hat{a}_q})$ for $q = 0,\cdots,p$. Hence $(\mathfrak{x}', \mathfrak{v}')$, $(\mathfrak{y}, \mathfrak{w})$ is a double representation of order $j-1$ of (x,v,w) with

$$\mathfrak{x}'_j = h_0\mathfrak{y}_0 + \cdots + h_{p-j}\mathfrak{y}_{p-j} .$$

Claim 1 is proved and $h_q = 0$ for $q = p-j+1,\cdots,p$ shall be assumed.

CLAIM 2. W.l.o.g. $h_{p-j} \neq 0$ can be assumed.

Proof of Claim 2. Assume that $h_{p-j} = 0$. Take $\varepsilon > 0$. Define $\mathfrak{x}'_q = \mathfrak{x}_q$ if $q \neq j$. Define

$$\mathfrak{x}'_j = \sum_{q=0}^{p-j-1} h_q\mathfrak{y}_q + \varepsilon\mathfrak{y}_{p-j} .$$

Define $\mathfrak{v}'_\mu = \mathfrak{v}_\mu$ if $\mu \neq \hat{a}_j$ and $\mathfrak{v}'_{\hat{a}_j} = \mathfrak{x}'_j$. By continuity, $\varepsilon_1 > 0$ exists such that $\mathfrak{x}'_0 \wedge \cdots \wedge \mathfrak{x}'_p \neq 0$ and $\mathfrak{v}'_0 \wedge \cdots \wedge \mathfrak{v}'_{\hat{a}_p} \neq 0$ if $0 < \varepsilon \leq \varepsilon_1$. Then $x = P(\mathfrak{x}'_0 \wedge \cdots \wedge \mathfrak{x}'_p)$. Define $v'_q = P(\mathfrak{v}'_0 \wedge \cdots \wedge \mathfrak{v}'_{\hat{a}_q})$ for $q = 0,\cdots,p$. Then $v' \in F(\mathfrak{a})$ and $(x,v',w) \in S(\mathfrak{a},\mathfrak{b})$. Also $(\mathfrak{x}', \mathfrak{v}')$ and $(\mathfrak{y}, \mathfrak{w})$ is a double representation of order $j-1$ of (x,v',w). Claim 2 is proved and $h_{p-j} \neq 0$ and $h_q = 0$ for $q = p-j+1,\cdots,p$ shall be assumed.

Define $\mathfrak{y}'_q = \mathfrak{y}_q$ if $q \neq p-j$ and $\mathfrak{y}'_{p-j} = \mathfrak{x}_j$. Then

$$\mathfrak{y}'_0 \wedge \cdots \wedge \mathfrak{y}'_p = (\mathfrak{y}_0 \wedge \cdots \wedge \mathfrak{y}_p)h_{p-j} \neq 0$$

and $x = \mathbf{P}(\mathfrak{y}'_0 \wedge \cdots \wedge \mathfrak{y}'_p)$. Define $\mathfrak{w}'_\mu = \mathfrak{w}_\mu$ if $\mu \neq \hat{b}_{p-j}$ and $\mathfrak{w}'_\mu = \mathfrak{x}_j$ if $\mu = \hat{b}_{p-j}$. If $q \geq p-j$, then

$$\mathfrak{w}'_0 \wedge \cdots \wedge \mathfrak{w}'_{\hat{b}_q} = \mathfrak{w}_0 \wedge \cdots \wedge \mathfrak{w}_{\hat{b}_q} h_{p-j} \neq 0 .$$

Therefore $w_q = \mathbf{P}(\mathfrak{w}'_0 \wedge \cdots \wedge \mathfrak{w}'_{\hat{b}_q})$ for all $q = 0, \cdots, p$. Define $\mathfrak{y}' = (\mathfrak{y}'_0, \cdots, \mathfrak{y}'_p)$ and $\mathfrak{w}' = (\mathfrak{w}'_0, \cdots, \mathfrak{w}'_{\hat{b}_p})$. Then $(\mathfrak{y}', \mathfrak{w}')$ is a representation of (x,w). Also $(\mathfrak{x}, \mathfrak{v})$ and $(\mathfrak{y}', \mathfrak{w}')$ is a double representation of order j of (x,v,w). Therefore $H(j) \neq \emptyset$. By induction $H(\mathfrak{a}, \mathfrak{b}) = H(p) \neq \emptyset$. Because $S(\mathfrak{a}, \mathfrak{b})$ is irreducible the Zariski open subset $H(\mathfrak{a}, \mathfrak{b}) \neq \emptyset$ is dense in $S(\mathfrak{a}, \mathfrak{b})$; q.e.d.

The concept of a modification was explained in the proof of Proposition 5.4. The concept of a rank of a holomorphic map and the properties of the rank are given in [1]. Because $S(\mathfrak{a}, \mathfrak{b})$ is irreducible, the analytic set $T(\mathfrak{a}, \mathfrak{b}) = \xi(S(\mathfrak{a}, \mathfrak{b}))$ is irreducible with $\dim T(\mathfrak{a}, \mathfrak{b}) = \operatorname{rank} \xi$.

THEOREM 6.8. *In diagram 6.1 assume that* $\operatorname{rank} \xi = \dim S(\mathfrak{a}, \mathfrak{b})$. *Then* $\mathfrak{b}^* \geq \mathfrak{a}$ *and* $\xi : S(\mathfrak{a}, \mathfrak{b}) \to T(\mathfrak{a}, \mathfrak{b})$ *is a modification.*

Proof. Recall $d(\mathfrak{a}, \mathfrak{b}) = \dim S(\mathfrak{a}, \mathfrak{b})$. The analytic set

$$E_0 = \{z \in S(\mathfrak{a}, \mathfrak{b}) \mid \operatorname{rank}_z \xi < d(\mathfrak{a}, \mathfrak{b})\}$$

is thin in $S(\mathfrak{a}, \mathfrak{b})$. Let Σ_0 and Σ_1 be the sets of singular points of $S(\mathfrak{a}, \mathfrak{b})$ respectively $T(\mathfrak{a}, \mathfrak{b})$. Define $E_1 = S(\mathfrak{a}, \mathfrak{b}) - H(\mathfrak{a}, \mathfrak{b})$. Then $E_2 = E_0 \cup E_1 \cup \Sigma_0$ is thin analytic in $S(\mathfrak{a}, \mathfrak{b})$. Hence $D = \xi(E_2) \cup \Sigma_1$ is a thin analytic subset of $T(\mathfrak{a}, \mathfrak{b})$. Then $E = \xi^{-1}(D)$ is thin analytic in $S(\mathfrak{a}, \mathfrak{b})$ with $E \supseteq E_2$. Define $N = S(\mathfrak{a}, \mathfrak{b}) - E$ and $M = T(\mathfrak{a}, \mathfrak{b}) - D$. Then N and M are connected, complex manifolds of dimension $d(\mathfrak{a}, \mathfrak{b})$ and $\xi : N \to M$ is a proper, surjective, light holomorphic map.

CLAIM 1. If $(v,w) \in M$, then $\dim E(v_q) \cap E(w_{p-q}) = 1$ for $q = 0,1,\cdots,p$.

Proof of Claim 1. Take $x \in G_p(V)$ such that $(x,v,w) \in N$. Then $x \in H(\mathfrak{a},\mathfrak{b})$. Hence

$$1 = \dim E(x) \cap E(v_q) \cap E(w_{p-q}) \leq \dim E(v_q) \cap E(w_{p-q})$$

for $q = 0,\cdots,p$. Assume $q \in \mathbb{Z}[0,p]$ exists such that $\dim E(v_q) \cap E(w_{p-q}) > 1$. Then $\mathfrak{z} \in E(v_q) \cap E(w_{p-q}) - E(x)$ exists. Let $(\mathfrak{x},\mathfrak{v},\mathfrak{w})$ be a representation of (x,v,w). Then $\mathfrak{x}'_q = \mathfrak{x}_q + \lambda\mathfrak{z}$ does not belong to $E(x)$ if $0 \neq \lambda \in \mathbb{C}$. Define $\mathfrak{x}'_\mu = \mathfrak{x}_\mu$ for $\mu \neq q$. Then $r > 0$ exists such that $\mathfrak{x}'_0 \wedge \cdots \wedge \mathfrak{x}'_p \neq 0$ if $|\lambda| < r$. If $|\lambda| < r$, define $x'(\lambda) = \mathbb{P}(\mathfrak{x}'_0 \wedge \cdots \wedge \mathfrak{x}'_p)$. Then $x \neq x'(\lambda)$ for $\lambda \neq 0$ and $x'(\lambda) \to x$ for $\lambda \to 0$. Also $\mathfrak{x}'_0,\cdots,\mathfrak{x}'_\mu$ belong to $E(x'(\lambda)) \cap E(v_\mu)$ for $\mu = 0,\cdots,p$. Hence $(x'(\lambda),v) \in S(\mathfrak{a})$. Also $\mathfrak{x}'_p,\cdots,\mathfrak{x}'_{p-\mu}$ belong to $E(x'(\lambda)) \cap E(w_\mu)$ for $\mu = 0,\cdots,p$. Hence $(x'(\lambda),w) \in S(\mathfrak{b})$. Therefore $(x'(\lambda),v,w) \in S(\mathfrak{a},\mathfrak{b})$, which implies $(x'(\lambda),v,w) \in \xi^{-1}(v,w)$. Because $\xi^{-1}(v,w)$ is zero dimensional, $x'(\lambda)$ is constant. Hence $x'(\lambda) = x$ for all $|\lambda| < r$, which is wrong. Claim 1 is proved.

Take $(v,w) \in M$. If $0 \leq q \leq p$, then

$$\begin{aligned} 1 &= \dim E(v_q) \cap E(w_{p-q}) \geq \dim E(v_q) + \dim E(w_{p-q}) - \dim V \\ &= a_q + q + 1 + b_{p-q} + p - q + 1 - n - 1 = a_q - (n - p - b_{p-q}) + 1 . \end{aligned}$$

Hence $n - p - b_{p-q} \geq a_q$ which means $\mathfrak{b}^* \geq \mathfrak{a}$.

CLAIM 2. The map $\xi : N \to M$ is biholomorphic.

Proof of Claim 2. Take $(v,w) \in M$. Assume (x,v,w) and (x',v,w) belong to $\xi^{-1}(v,w)$. Claim 1 implies that

$$L_q(x,v,w) = E(x) \cap E(v_q) \cap E(w_{p-q}) = E(v_q) \cap E(w_{p-q})$$

$$L_q(x',v,w) = E(x') \cap E(v_q) \cap E(w_{p-q}) = E(v_q) \cap E(w_{p-q})$$

for $q = 0,\cdots,p$. Hence $L_q(x,v,w) = L_q(x',v,w)$. Therefore

$$E(x) = \bigoplus_{q=0}^{p} L_q(x,v,w) = \bigoplus_{q=0}^{p} L_q(x',v,w) = E(x') .$$

which implies that $x = x'$. The map $\xi : N \to M$ is injective, hence biholomorphic. Claim 2 is proved.

By claim 2, $\xi : S(\mathfrak{a},\mathfrak{b}) \to T(\mathfrak{a},\mathfrak{b})$ is a modification, q.e.d.

THEOREM 6.9. *Take* $\mathfrak{a}$ *and* $\mathfrak{b}$ *in* $\mathfrak{S}(p,n)$ *with* $\vec{\mathfrak{a}} + \vec{\mathfrak{b}} = d(p,n)$. *Then* $\operatorname{rank}\xi = \dim S(\mathfrak{a},\mathfrak{b})$ *if and only if* $\mathfrak{b}^* = \mathfrak{a}$ *(Diagram 6.1). If* $\mathfrak{b}^* = \mathfrak{a}$, *then* $T(\mathfrak{a},\mathfrak{b}) = F(\mathfrak{a}) \times F(\mathfrak{b})$.

Proof. By Lemma 6.1 $\dim S(\mathfrak{a},\mathfrak{b}) = d(\mathfrak{a},\mathfrak{b}) = d(\mathfrak{a}) + d(\mathfrak{b})$ is the dimension of $F(\mathfrak{a}) \times F(\mathfrak{b})$. Hence if $\operatorname{rank}\xi = d(\mathfrak{a},\mathfrak{b})$, then $T(\mathfrak{a},\mathfrak{b})$ is an analytic subset of $F(\mathfrak{a}) \times F(\mathfrak{b})$ with the dimension of $F(\mathfrak{a}) \times F(\mathfrak{b})$. Since $F(\mathfrak{a}) \times F(\mathfrak{b})$ is irreducible, we have $T(\mathfrak{a},\mathfrak{b}) = F(\mathfrak{a}) \times F(\mathfrak{b})$. Lemma 6.3 implies that $\mathfrak{a} \geq \mathfrak{b}^*$. Theorem 6.8 implies that $\mathfrak{b}^* \geq \mathfrak{a}$. Hence $\mathfrak{b}^* = \mathfrak{a}$.

Assume that $\mathfrak{b}^* = \mathfrak{a}$. Let $e_0,\cdots,e_n$ be a base of V. Define $\mathfrak{v}_\mu = e_\mu$ for $\mu = 0,\cdots,\hat{a}_p$ and $\mathfrak{w}_\mu = e_{n-\mu}$ for $\mu = 0,\cdots,\hat{b}_p$. Define

$$v_q = P(\mathfrak{v}_0 \wedge \cdots \wedge \mathfrak{v}_{\hat{a}_q}) \qquad w_q = P(\mathfrak{w}_0 \wedge \cdots \wedge \mathfrak{w}_{\hat{b}_q})$$

for $q = 0,\cdots,p$. Then $v = (v_0,\cdots,v_p) \in F(\mathfrak{a})$ and $w = (w_0,\cdots,w_p) \in F(\mathfrak{b})$. Define $\mathfrak{x}_q = e_{\hat{a}_q}$ for $q = 0,\cdots,p$ and $x = P(\mathfrak{x}_0 \wedge \cdots \wedge \mathfrak{x}_p)$. The vectors $\mathfrak{x}_0,\cdots,\mathfrak{x}_q$ span $E(x) \cap E(v_q)$. Hence $(x,v) \in S(\mathfrak{a})$. Observe $\hat{b}_q = b_q + q = n - p - a_{p-q} + q = n - \hat{a}_{p-q}$ and $e_{\hat{b}_q} = e_{\hat{a}_{p-q}} = \mathfrak{x}_{p-q}$. Therefore $\mathfrak{x}_p, \mathfrak{x}_{p-1},\cdots,\mathfrak{x}_{p-q}$ span $E(x) \cap E(w_q)$ for $q = 0,\cdots,p$. Hence $(x,w) \in S(\mathfrak{b})$ and $z = (x,v,w) \in S(\mathfrak{a},\mathfrak{b})$. The vectors $e_0,\cdots,e_{\hat{a}_q}$ span $E(v_q)$

and the vectors $e_n, e_{n-1}, \cdots, e_{\hat{a}_q}$ span $E(w_{p-q})$. Therefore $E(v_q) \cap E(w_{p-q}) = \mathbb{C}\, e_{\hat{a}_q} = \mathbb{C}\, \mathfrak{x}_q$. Take $(x', v, w) \in S(\mathfrak{a}, \mathfrak{b})$. Then $\dim E(x') \cap E(v_q) \geq q+1$ and $\dim E(x') \cap E(w_{p-q}) \geq p-q+1$ imply

$$\dim E(x') \cap E(v_q) \cap E(w_{p-q}) \geq 1 .$$

Therefore $E(x') \cap E(v_q) \cap E(w_{p-q}) = E(v_q) \cap E(w_{p-q}) = \mathbb{C}\, \mathfrak{x}_q$. Hence $\mathfrak{x}_q \in E(x')$ for $q = 0, \cdots, p$, which implies that $E(x) \subseteq E(x')$. Since $E(x)$ and $E(x')$ have the same dimension, $E(x) = E(x')$ and $x = x'$ follow. Hence $\xi^{-1}(v,w) = \{z\}$ is zero dimensional. Consequently,

$$\begin{aligned} \operatorname{rank} \xi \geq \operatorname{rank}_z \xi &= \dim S(\mathfrak{a}, \mathfrak{b}) - \dim_z \xi^{-1}(v,w) \\ &= \dim S(\mathfrak{a}, \mathfrak{b}) \geq \operatorname{rank} \xi . \end{aligned}$$

Hence $\operatorname{rank} \xi = \dim S(\mathfrak{a}, \mathfrak{b})$; q.e.d.

The basic geometric results having been obtained, we can apply the calculus of exterior forms.

c) *The Duality Theorem*

THEOREM 6.10 (Ehresmann [11]). *Take* $\mathfrak{a}$ *and* $\mathfrak{b}$ *in* $\mathfrak{S}(p,n)$ *with* $\vec{\mathfrak{a}} + \vec{\mathfrak{b}} = d(p,n)$. *If* $\mathfrak{a} \neq \mathfrak{b}^*$, *then* $c(\mathfrak{a}) \cap c(\mathfrak{b}) = 0$ *if* $\mathfrak{a} = \mathfrak{b}^*$, *then* $c(\mathfrak{a}) \wedge c(\mathfrak{b}) = c_{n-p}[p]^{p+1}$. *In particular*

$$\int_{G_p(V)} c(\mathfrak{a}) \wedge c(\mathfrak{b}) = \begin{cases} 0 & \text{if} \quad \mathfrak{a} \neq \mathfrak{b}^* \\ 1 & \text{if} \quad \mathfrak{a} = \mathfrak{b}^* . \end{cases}$$

Proof. Consider diagram 6.1. Then $\Omega = \alpha_{\mathfrak{a}}^*(\Omega_{\mathfrak{a}}) \wedge \alpha_{\mathfrak{b}}^*(\Omega_{\mathfrak{b}})$ has degree $2(d(\mathfrak{a}) + d(\mathfrak{b}))$. If $\mathfrak{a} \neq \mathfrak{b}^*$, then $\operatorname{rank} \xi < \dim S(\mathfrak{a}, \mathfrak{b})$ by Theorem 6.9, where $\dim S(\mathfrak{a}, \mathfrak{b}) = d(\mathfrak{a}) + d(\mathfrak{b})$ by Lemma 6.1. Hence $\xi^*(\Omega) = 0$. Proposition 6.2 implies $c(\mathfrak{a}) \wedge c(\mathfrak{b}) = \rho_* \xi^*(\Omega) = 0$. Assume that $\mathfrak{a} = \mathfrak{b}^*$.

Then $\operatorname{rank} \xi = \dim S(\mathfrak{a}, \mathfrak{b})$ and $\xi : S(\mathfrak{a}, \mathfrak{b}) \to F(\mathfrak{a}) \times F(\mathfrak{b})$ is a modification, by Theorem 6.9 and Theorem 6.8. Therefore

$$\int_{G_p(V)} c(\mathfrak{a}) \wedge c(\mathfrak{b}) = \int_{G_p(V)} \rho_* \xi^*(\Omega) = \int_{S(\mathfrak{a}, \mathfrak{b})} \xi^*(\Omega)$$

$$= \int_{F(\mathfrak{a}) \times F(\mathfrak{b})} \alpha_{\mathfrak{a}}^*(\Omega_{\mathfrak{a}}) \wedge \alpha_{\mathfrak{b}}^*(\Omega_{\mathfrak{b}})$$

$$= \int_{F(\mathfrak{a})} \Omega_{\mathfrak{a}} \int_{F(\mathfrak{b})} \Omega_{\mathfrak{b}} = 1 .$$

Because $c_{n-p}[p]^{p+1} > 0$, a function γ exists such that

$$c(\mathfrak{a}) \wedge c(\mathfrak{b}) = \gamma c_{n-p}[p]^{p+1} .$$

Because $c(\mathfrak{a})$, $c(\mathfrak{b})$ and $c_{n-p}[p]$ are invariant under the unitary group, γ is constant. Integration over $G_p(V)$ shows that $\gamma = 1$ where Theorem 4.4 was used; q.e.d.

THEOREM 6.11. *Take* $p \in \mathbb{Z}[0,n]$ *and* $m \in \mathbb{Z}[0, d(p,n)]$. *Define* $s = d(p,n) - m$. *Define* $\mathfrak{S}(p,n,m) = \{\mathfrak{a} \in \mathfrak{S}(p,n) \mid \vec{\mathfrak{a}} = m\}$. *Then the family* $\{c(\mathfrak{a})\}_{\mathfrak{a} \in \mathfrak{S}(p,n,m)}$ *is a base of the vector space* $\operatorname{Inv}^{2s}(G_p(V))$ *and therefore a cohomology base over* $\mathbb{C}$ *for dimension* $2s$.

Proof. By (2.11) and (2.12) $\dim \operatorname{Inv}^{2s}(G_p(V)) = \# \mathfrak{S}(p,n,m)$. Hence it suffices to show that the $c(\mathfrak{a})$ are linear independent. Assume that for each $\mathfrak{a} \in \mathfrak{S}(p,n,m)$ there exists a constant $\gamma_{\mathfrak{a}} \in \mathbb{C}$ such that

$$\sum_{\mathfrak{a} \in \mathfrak{S}(p,n,m)} \gamma_{\mathfrak{a}} c(\mathfrak{a}) = 0 .$$

Take $\mathfrak{b} \in \mathfrak{S}(p,n,m)$. Then

$$\gamma_{\mathfrak{b}} = \int_{G_p(V)} \sum_{\mathfrak{a} \in \mathfrak{S}(p,n,m)} \gamma_{\mathfrak{a}} c(\mathfrak{a}) \wedge c(\mathfrak{b}^*) = 0 ; \qquad \text{q.e.d.}$$

Take $\mathfrak{a} \in \mathfrak{S}(p,n)$ and $v \in F(\mathfrak{a})$. Then Theorem 5.2 and Theorem 6.10 imply that

$$\int_{S(v,\mathfrak{a})} c(\mathfrak{b}) = \begin{cases} 0 & \text{if } \mathfrak{b} \neq \mathfrak{a}^* \\ 1 & \text{if } \mathfrak{b} = \mathfrak{a}^* \end{cases} \quad \text{and} \quad \mathfrak{b} \in \mathfrak{S}(p,n,\vec{\mathfrak{a}}) .$$

Hence $c(\mathfrak{a}^*)$ is the dual to $S(v,\mathfrak{a})$. Recall that $c(\mathfrak{a})$ is the Poincaré dual of $S(v,\mathfrak{a})$.

d) *Matsushima's Theorem*

Take $\mathfrak{a} \in \mathfrak{S}(p,n)$ and $r \in \mathbf{Z}$ with $0 \leq r \leq d(p,n) - \vec{\mathfrak{a}}$. Define

$$\Delta(\mathfrak{a}, r) = \{\mathfrak{b} \in \mathfrak{S}(p,n) \mid \mathfrak{b} \geq \mathfrak{a} \text{ and } \vec{\mathfrak{b}} = \vec{\mathfrak{a}} + r\} .$$

THEOREM 6.12 (Matsushima [21]). *Take* $p \in \mathbf{Z}[0,n]$ *and* $\mathfrak{a} \in \mathfrak{S}(p,n)$. *Let* r *be an integer with* $0 \leq r \leq d(p,n) - \vec{\mathfrak{a}}$ *and with* $s = d(\mathfrak{a}) - r \geq 0$. *Let* λ *be a form of class* C^∞ *and degree* $2s$ *on* $F(\mathfrak{a})$ *such that* λ *is invariant under the action of the unitary group* $\mathfrak{U}(V)$ *on* $F(\mathfrak{a})$. *Let* $\sigma : S(\mathfrak{a}) \to F(\mathfrak{a})$ *and* $\pi : S(\mathfrak{a}) \to G_p(V)$ *be the projections. Define* $\Lambda = \pi_* \sigma^*(\lambda)$. *Then there exists one and only one* $\gamma_{\mathfrak{c}} \in \mathbf{C}$ *for each* $\mathfrak{c} \in \Delta(\mathfrak{a}, r)$ *such that*

$$\Lambda = \sum_{\mathfrak{c} \in \Delta(\mathfrak{a}, r)} \gamma_{\mathfrak{c}} c(\mathfrak{c}) .$$

Moreover, if λ *is real, each* $\gamma_{\mathfrak{c}}$ *is real, and if* $\lambda \geq 0$, *then* $\gamma_{\mathfrak{c}} \geq 0$ *for each* $\mathfrak{c} \in \Delta(\mathfrak{a}, r)$.

Proof. Define $m = \vec{\mathfrak{a}} + r$. Since π has fiber dimension $d(\mathfrak{a}) + \vec{\mathfrak{a}} - d(p,n)$. The form Λ has degree $2(d(p,n) - m)$. Take $g \in \mathfrak{U}(V)$. Then $g^*(\Lambda) = g^* \pi_* \sigma^*(\lambda) = \pi_* \sigma^*(g^* \lambda) = \pi_* \sigma^*(\lambda) = \Lambda$. Unique numbers $\gamma_{\mathfrak{c}} \in \mathbb{C}$ exist such that

$$\Lambda = \sum_{\mathfrak{c} \in \mathfrak{S}(p,n,m)} \gamma_{\mathfrak{c}} \, c(\mathfrak{c}) \, . \tag{6.6}$$

Take $\mathfrak{c} \in \mathfrak{S}(p,n,m)$. Define $\mathfrak{b} = \mathfrak{c}^*$. Consider diagram 6.1 with $\pi = \pi_{\mathfrak{a}}$ and $\sigma = \sigma_{\mathfrak{a}}$. Theorem 6.10 and Theorem 4.1 imply that

$$\gamma_{\mathfrak{c}} = \int_{G_p(V)} \Lambda \wedge c(\mathfrak{c}^*)$$

$$= \int_{G_p(V)} \pi_{\mathfrak{a}_*} \sigma_{\mathfrak{a}}^*(\lambda) \wedge \pi_{\mathfrak{b}_*} \sigma_{\mathfrak{b}}^*(\Omega_{\mathfrak{b}})$$

$$= \int_{G_p(V)} \rho_*(\beta_{\mathfrak{a}}^* \sigma_{\mathfrak{a}}^*(\lambda) \wedge \beta_{\mathfrak{b}}^* \sigma_{\mathfrak{b}}^*(\Omega_{\mathfrak{b}}))$$

$$= \int_{G_p(V)} \rho_* \xi^*(\alpha_{\mathfrak{a}}^*(\lambda) \wedge \alpha_{\mathfrak{b}}^*(\Omega_{\mathfrak{b}}))$$

$$= \int_{S(\mathfrak{a}, \mathfrak{b})} \xi^*(\alpha_{\mathfrak{a}}^*(\lambda) \wedge \alpha_{\mathfrak{b}}^*(\Omega_{\mathfrak{b}})) \, .$$

The analytic set $T(\mathfrak{a}, \mathfrak{b}) = \xi(S(\mathfrak{a}, \mathfrak{b}))$ is irreducible with $\dim T(\mathfrak{a}, \mathfrak{b}) = \operatorname{rank} \xi$. Let $\xi_0 = \xi : S(\mathfrak{a}, \mathfrak{b}) \to T(\mathfrak{a}, \mathfrak{b})$ be the restriction and let $\iota : T(\mathfrak{a}, \mathfrak{b}) \to F(\mathfrak{a}) \times F(\mathfrak{b})$ be the inclusion. Then $\xi = \iota \circ \xi_0$. The degree of $\overset{*}{a}_{\mathfrak{a}}(\lambda) \wedge \overset{*}{a}_{\mathfrak{b}}(\Omega_{\mathfrak{b}})$ is $2 \dim S(\mathfrak{a}, \mathfrak{b})$. Hence if $\operatorname{rank} \xi < \dim S(\mathfrak{a}, \mathfrak{b})$, then

$$\xi(\overset{*}{a}_{\mathfrak{a}}(\lambda) \wedge \overset{*}{a}_{\mathfrak{b}}(\Omega_{\mathfrak{b}})) = \xi_0^* \iota^*(\overset{*}{a}_{\mathfrak{a}}(\lambda) \wedge \overset{*}{a}_{\mathfrak{b}}(\Omega_{\mathfrak{b}})) = \xi_0^*(0) = 0 .$$

Therefore $\gamma_{\mathfrak{c}} = 0$. Hence if $\gamma_{\mathfrak{c}} \neq 0$, then $\operatorname{rank} \xi = \dim S(\mathfrak{a}, \mathfrak{b})$. Theorem 6.8 implies $\mathfrak{c} = \mathfrak{b}^* \geq \mathfrak{a}$. Therefore $\mathfrak{c} \in \Delta(\mathfrak{a}, r)$ and (6.6) implies (6.5). Observe that

$$\gamma_{\mathfrak{c}} = \int_{S(\mathfrak{a}, \mathfrak{b})} \xi^*(\overset{*}{a}_{\mathfrak{a}}(\lambda) \wedge \overset{*}{a}_{\mathfrak{b}}(\Omega_{\mathfrak{b}})) . \tag{6.7}$$

Hence if λ is real, then $\gamma_{\mathfrak{c}}$ is real, if $\lambda \geq 0$, then $\gamma_{\mathfrak{c}} \geq 0$; q.e.d.

REMARK. If $\gamma_{\mathfrak{c}} \neq 0$, then $\operatorname{rank} \xi = \dim S(\mathfrak{a}, \mathfrak{b})$ and ξ_0 is a modification. Therefore

$$\gamma_{\mathfrak{c}} = \int_{T(\mathfrak{a}, \mathfrak{b})} \overset{*}{a}_{\mathfrak{a}}(\lambda) \wedge \overset{*}{a}_{\mathfrak{b}}(\Omega_{\mathfrak{b}}) . \tag{6.8}$$

with $\mathfrak{b} = \mathfrak{c}^*$. Therefore $\gamma_{\mathfrak{c}}$ can be expressed directly by λ and $\mathfrak{b}$.

Matsushima [21] gave a totally different proof which uses Lie algebra and difficult results of Kostant. After this paper was written in its final form, but before it was sent to the publisher, Matsushima received a letter from James Damon in which Damon proves Theorem 6.12 using Damon [9] and [10]. His method employs the Gysin homomorphism computed by a residue calculus. Thus Damon uses deep results in cohomology theory. The proof provided here rests on fiber integration for fibers with singularities. This operator is easy to understand but difficult to construct (Tung [33]). Once this operator is accepted and once the elementary but tedious geometric considerations Lemma 6.3 - Theorem 6.9 are settled, the proof of Matsushima's Theorem becomes a triviality and the same remark applies to the Duality Theorem and to Pieri's Theorem, where the latter requires some additional geometric inquiries.

7. THE THEOREMS OF PIERI AND GIAMBELLI

The theorem of Giambelli can be reduced to the theorem of Pieri by elementary considerations. Both theorems have been proved on several occasions; see Giambelli [13], Hodge [18], Chern [5] and Vesentini [34]. Here fiber integration and the triple intersection diagram will give a proof of Pieri's Theorem.

a) *Triple intersection*

Take $p \in Z[0,n]$ and $\mathfrak{a}, \mathfrak{b}, \mathfrak{c}$ in $\mathfrak{S}(p,n)$. Define $S(\mathfrak{a}, \mathfrak{b}, \mathfrak{c})$ as the set of all $(x,v,w,z) \in G_p(V) \times F(\mathfrak{a}) \times F(\mathfrak{b}) \times F(\mathfrak{c})$ such that $(x,v) \in S(\mathfrak{a})$, $(x,w) \in S(\mathfrak{b})$ and $(x,z) \in S(\mathfrak{c})$. Obviously, $S(\mathfrak{a}, \mathfrak{b}, \mathfrak{c})$ is an analytic subset. Consider the *triple intersection diagram* 7.1. The spaces are enumerated by

$0 = G_p(V)$

$1 = S(\mathfrak{a}, \mathfrak{b}, \mathfrak{c})$	$8 = F(\mathfrak{a}) \times F(\mathfrak{b}) \times F(\mathfrak{c})$
$2 = S(\mathfrak{b}, \mathfrak{c})$	$9 = F(\mathfrak{b}) \times F(\mathfrak{c})$
$3 = S(\mathfrak{c})$	$10 = F(\mathfrak{c})$
$4 = S(\mathfrak{a}, \mathfrak{c})$	$11 = F(\mathfrak{a}) \times F(\mathfrak{c})$
$5 = S(\mathfrak{a})$	$12 = F(\mathfrak{a})$
$6 = S(\mathfrak{a}, \mathfrak{b})$	$13 = F(\mathfrak{a}) \times F(\mathfrak{b})$
$7 = S(\mathfrak{b})$	$14 = F(\mathfrak{b})$

A map from a space numbered x into a space numbered y is denoted by $[y,x]$. All the maps in the diagram are projections. The diagram commutes.

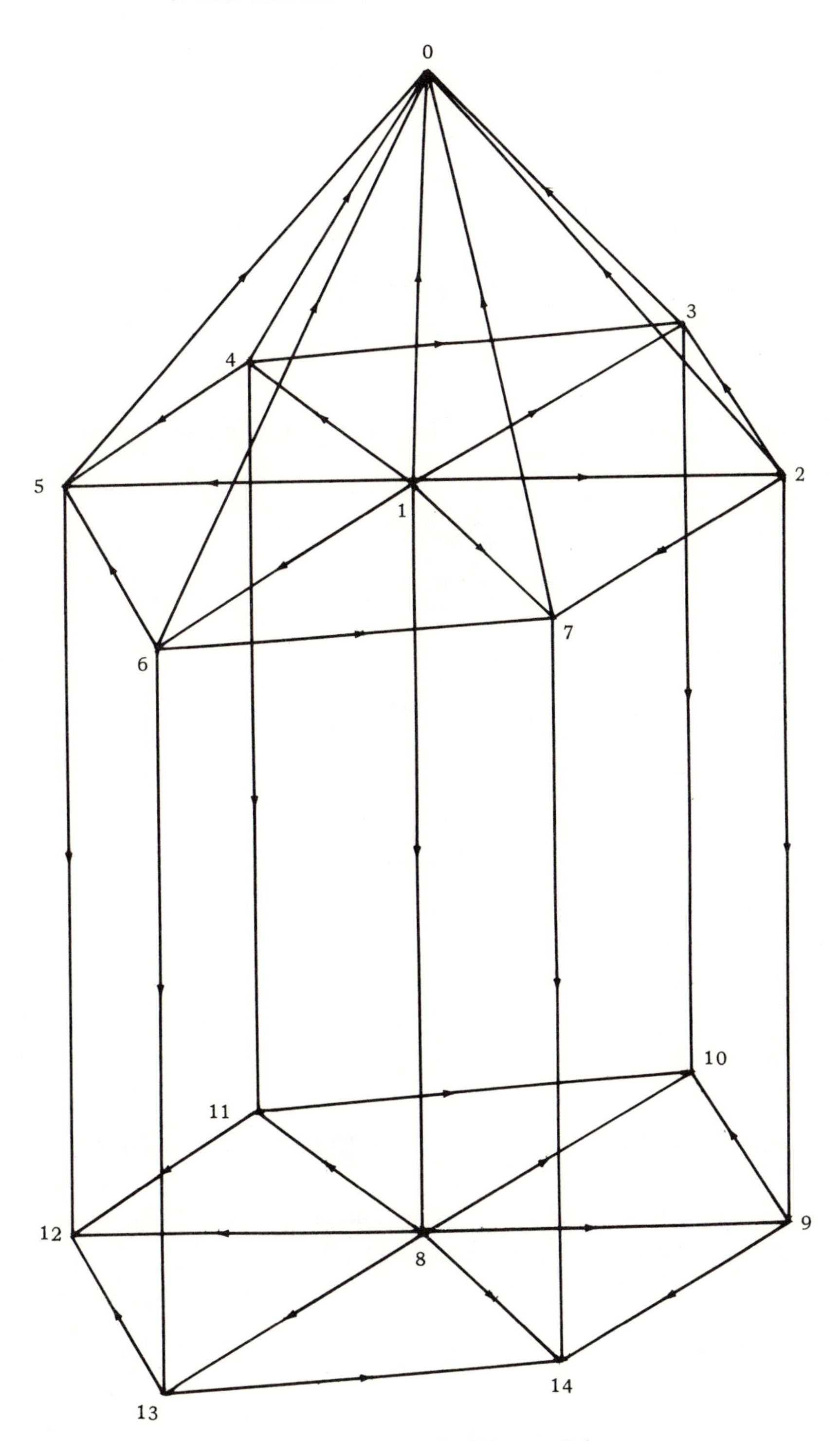

Triple Intersection Diagram 7.1

Also define

$$\zeta = [8,1] : S(\mathfrak{a},\mathfrak{b},\mathfrak{c}) \to F(\mathfrak{a}) \times F(\mathfrak{b}) \times F(\mathfrak{c})$$

$$\eta = [0,1] : S(\mathfrak{a},\mathfrak{b},\mathfrak{c}) \to G_p(V)$$

$$\theta_{\mathfrak{a}} = [12,8] : F(\mathfrak{a}) \times F(\mathfrak{b}) \times F(\mathfrak{c}) \to F(\mathfrak{a})$$

$$\theta_{\mathfrak{b}} = [14,8] : F(\mathfrak{a}) \times F(\mathfrak{b}) \times F(\mathfrak{c}) \to F(\mathfrak{b})$$

$$\theta_{\mathfrak{c}} = [10,8] : F(\mathfrak{a}) \times F(\mathfrak{b}) \times F(\mathfrak{c}) \to F(\mathfrak{c})$$

LEMMA 7.1. $S(\mathfrak{a},\mathfrak{b},\mathfrak{c})$ *is irreducible. Let* $d(\mathfrak{a},\mathfrak{b},\mathfrak{c})$ *be the dimension of* $S(\mathfrak{a},\mathfrak{b},\mathfrak{c})$ *and define*

$$\Omega = \theta_{\mathfrak{a}}^*(\Omega_{\mathfrak{a}}) \wedge \theta_{\mathfrak{b}}^*(\Omega_{\mathfrak{b}}) \wedge \theta_{\mathfrak{c}}^*(\Omega_{\mathfrak{c}}) \tag{7.2}$$

Then

$$d(\mathfrak{a},\mathfrak{b},\mathfrak{c}) = d(\mathfrak{a}) + d(\mathfrak{b}) + d(\mathfrak{c}) + \vec{\mathfrak{a}} + \vec{\mathfrak{b}} + \vec{\mathfrak{c}} - 2d(p,n) \tag{7.3}$$

$$c(\mathfrak{a}) \wedge c(\mathfrak{b}) \wedge c(\mathfrak{c}) = \eta_* \zeta^*(\Omega) . \tag{7.4}$$

Proof. Consider the subdiagram

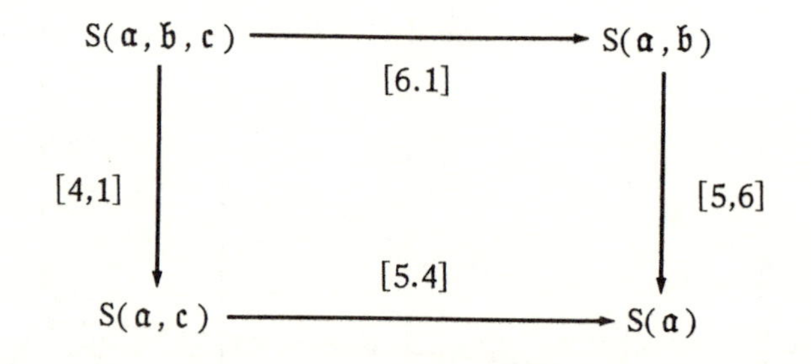

$$S(\mathfrak{a},\mathfrak{b},\mathfrak{c}) = \{(x,v,w,z) \mid (x,v,w) \in S(\mathfrak{a},\mathfrak{b}) \text{ and } (x,v,z) \in S(\mathfrak{a},\mathfrak{c})\}$$

is the relative product of the maps [5,4] and [5.6]. Hence $[4,1] : S(\mathfrak{a},\mathfrak{b},\mathfrak{c}) \to S(\mathfrak{a},\mathfrak{c})$ is a holomorphic fiber bundle with

$$\begin{aligned} \dim S(\mathfrak{a},\mathfrak{b},\mathfrak{c}) &= \dim S(\mathfrak{a},\mathfrak{c}) + \text{fiber dim } [4,1] \\ &= \dim S(\mathfrak{a},\mathfrak{c}) + \text{fiber dim } [5,6] \\ &= d(\mathfrak{a},\mathfrak{c}) + d(\mathfrak{a},\mathfrak{b}) - d(\mathfrak{a}) - \vec{\mathfrak{a}} \\ &= d(\mathfrak{a}) + d(\mathfrak{b}) + d(\mathfrak{c}) + \vec{\mathfrak{a}} + \vec{\mathfrak{b}} + \vec{\mathfrak{c}} - 2d(p,n) \end{aligned}$$

by Lemma 6.1 and Theorem 2.6. By Lemma 6.1, the fibers of [5,6] are irreducible. Hence the fibers of [4,1] are irreducible. Since $S(\mathfrak{a},\mathfrak{c})$ is irreducible (Lemma 6.1), $S(\mathfrak{a},\mathfrak{b},\mathfrak{c})$ is irreducible by Lemma 1.2. Only (7.4) remains to be proved.

The double intersection subdiagram for $S(\mathfrak{a},\mathfrak{b})$ yields

$$c(\mathfrak{a}) \wedge c(\mathfrak{b}) = [0,6]_*[13,6]^*([12,13]^*(\Omega_\mathfrak{a}) \wedge [14,13]^*(\Omega_\mathfrak{b}))$$

Moreover $c(\mathfrak{c}) = [0,3]_*[10,3]^*(\Omega_\mathfrak{c})$. Consider the diagram 7.5

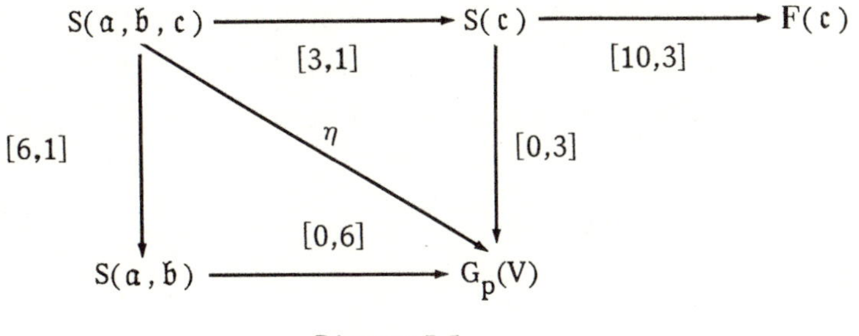

Diagram 7.5

Obviously $(S(\mathfrak{a},\mathfrak{b},\mathfrak{c}),\ [3,1],[6,1])$ is the relative product of [0,6] and [0,3]. Therefore Theorem 4.1 implies that

$$c(\mathfrak{a}) \wedge c(\mathfrak{b}) \wedge c(\mathfrak{c}) =$$

$$\begin{aligned} &= \eta_*([6,1]^*[13,6]^*([12,13]^*(\Omega_\mathfrak{a}) \wedge [14,13]^*(\Omega_\mathfrak{b})) \wedge [3,1]^*[10,3]^*(\Omega_\mathfrak{c})) \\ &= \eta_*(([12,13][13,6][6,1])^*(\Omega_\mathfrak{a}) \wedge ([14,13][13,6][6,1])^*(\Omega_\mathfrak{b}) \\ &\qquad \wedge ([10,3][3,1])^*(\Omega_\mathfrak{c})) \\ &= \eta_*(([12,8][8,1])^*(\Omega_\mathfrak{a}) \wedge ([14,8][8,1])^*(\Omega_\mathfrak{b}) \wedge ([10,8][8,1])^*(\Omega_\mathfrak{c})) \\ &= \eta_*\zeta^*(\theta_\mathfrak{a}^*(\Omega_\mathfrak{a}) \wedge \theta_\mathfrak{b}^*(\Omega_\mathfrak{b}) \wedge \theta_\mathfrak{c}^*(\Omega_\mathfrak{c})) = \eta_*\zeta^*(\Omega); \end{aligned}$$

q.e.d.

b) *The Theorem of Pieri*

LEMMA 7.2. *Take* $\mathfrak{a}$ *and* $\mathfrak{b}$ *in* $\mathfrak{S}(p,n)$. *Take* $h \in \mathbb{Z}[0,n-p]$. *Assume that* $\vec{\mathfrak{a}} + \vec{\mathfrak{b}} = d(p,n) + h$. *Define* $\mathfrak{c} = (n-p-h, n-p, \cdots, n-p)$. *Assume that the projection* $\zeta = [8,1]$ *is surjective. Then* ζ *is a modification and* $a_{j-1} \leq b_j^* \leq a_j$ *for* $j = 0, 1, \cdots, p$. *As before* $a_{-1} = 0$ *and* $b_j^* = n-p-b_{p-j}$.

Proof. Consider the triple intersection diagram 7.1. The maps [13,8] and [8,1] are surjective. Hence $[13,6] \circ [6,1]$ is surjective. Consequently [13,6] is surjective. Lemma 6.3 implies $\mathfrak{b}^* \leq \mathfrak{a}$. Hence $b_j^* \leq a_j$ for $j = 0, 1, \cdots, p$.

Since $\vec{\mathfrak{c}} = d(p,n) - h$, we have

$$\dim S(\mathfrak{a}, \mathfrak{b}, \mathfrak{c}) = d(\mathfrak{a}, \mathfrak{b}, \mathfrak{c}) = \dim F(\mathfrak{a}) \times F(\mathfrak{b}) \times F(\mathfrak{c}) .$$

Recall the definition of $H(\mathfrak{a}, \mathfrak{b})$ in 6.b). Since $A_0' = S(\mathfrak{a}, \mathfrak{b}) - H(\mathfrak{a}, \mathfrak{b})$ is thin analytic in $S(\mathfrak{a}, \mathfrak{b})$, also $\tilde{A}_0 = [6.1]^{-1}(A_0')$ is thin analytic in $S(\mathfrak{a}, \mathfrak{b}, \mathfrak{c})$ and $A_0 = \zeta(\tilde{A}_0)$ is thin analytic in $F(\mathfrak{a}) \times F(\mathfrak{b}) \times F(\mathfrak{c})$. An analytic subset A_1' of $F(\mathfrak{a}) \times F(\mathfrak{b})$ is defined by

$$A_1' = \bigcup_{q=0}^{p} \{(v,w) \in F(\mathfrak{a}) \times F(\mathfrak{b}) \mid \dim E(v_q) \cap E(w_{p-q}) \geq a_q - b_q^* + 2\} .$$

Let $\mathfrak{e}_0, \cdots, \mathfrak{e}_n$ be a base of V. Define $v_q = \mathbf{P}(\mathfrak{e}_0 \wedge \cdots \wedge \mathfrak{e}_{\hat{a}_q})$ in $G_{\hat{a}_q}(V)$ and $w_q = \mathbf{P}(\mathfrak{e}_n \wedge \cdots \wedge \mathfrak{e}_{n-\hat{b}_q})$ in $G_{\hat{b}_q}(V)$. Then $v = (v_0, \cdots, v_p) \in F(\mathfrak{a})$ and $w = (w_0, \cdots, w_p) \in F(\mathfrak{b})$ with $n - \hat{b}_{n-q} = b_q^* + q = \hat{b}_q^*$. Hence $w_{p-q} = \mathbf{P}(\mathfrak{e}_n \wedge \cdots \wedge \mathfrak{e}_{\hat{b}_q^*})$. Now $b_q^* \leq a_q$ implies $\hat{b}_q^* \leq \hat{a}_q$. Therefore $E(v_q) \cap E(w_{p-q})$ is spanned by $\mathfrak{e}_\mu$ for $\mu = \hat{b}_q^*, \cdots, \hat{a}_q$. Then $E(v_q) \cap E(w_{p-q})$ has dimension $a_q - b_q^* + 1$ for $q = 0, \cdots, p$. Hence $(v,w) \in F(\mathfrak{a}) \times F(\mathfrak{b}) - A_1'$. The analytic set A_1' is thin in $F(\mathfrak{a}) \times F(\mathfrak{b})$. Then $A_1 = A_1' \times F(\mathfrak{c})$ is thin analytic in $F(\mathfrak{a}) \times F(\mathfrak{b}) \times F(\mathfrak{c})$.

Because $S(\mathfrak{a}, \mathfrak{b}, \mathfrak{c})$ and $F(\mathfrak{a}) \times F(\mathfrak{b}) \times F(\mathfrak{c})$ are irreducible and have the same dimension, and because ζ is surjective, a thin analytic subset A_2 of $F(\mathfrak{a}) \times F(\mathfrak{b}) \times F(\mathfrak{c})$ exists such that

$$\zeta : S(\mathfrak{a}, \mathfrak{b}, \mathfrak{c}) - \zeta^{-1}(A_2) \to F(\mathfrak{a}) \times F(\mathfrak{b}) \times F(\mathfrak{c}) - A_2$$

is a light, proper, surjective, locally biholomorphic map. The analytic sets $A = A_0 \cup A_1 \cup A_2$ and $\widetilde{A} = \zeta^{-1}(A)$ are thin in $F(\mathfrak{a}) \times F(\mathfrak{b}) \times F(\mathfrak{c})$ respectively in $S(\mathfrak{a}, \mathfrak{b}, \mathfrak{c})$. The complements $N = F(\mathfrak{a}) \times F(\mathfrak{b}) \times F(\mathfrak{c}) - A$ and $G = S(\mathfrak{a}, \mathfrak{b}, \mathfrak{c}) - A$ are open, connected and dense in $F(\mathfrak{a}) \times F(\mathfrak{b}) \times F(\mathfrak{c})$ respectively in $S(\mathfrak{a}, \mathfrak{b}, \mathfrak{c})$. The map $\zeta_0 = \zeta : G \to N$ is proper, light, surjective and locally biholomorphic with $\zeta_0^{-1}(v,w,z) = \zeta^{-1}(v,w,z)$ if $(v,w,z) \in N$. The sheet number $s = \# \zeta_0^{-1}(v,w,z)$ is constant with $s \geq 1$. If $s = 1$ can be shown, then ζ is a modification.

Take $(v,w,z) \in N$. Take $q \in \mathbf{Z}[0,p]$. Then

$$a_q - b_q^* + 1 \geq \dim E(v_q) \cap E(w_{p-q}) \geq \hat{a}_q + 1 + \hat{b}_{p-q} + 1 - n - 1$$
$$= a_q + q + 1 + b_{p-q} + p - q - n = a_q - b_q^* + 1$$

$$\dim E(v_q) \cap E(w_{p-q}) = a_q - b_q^* + 1 \,. \tag{7.6}$$

Take $(x,v,w,z) \in G$. The definition of G implies that

$$\dim E(v_q) \cap E(x) = q+1 = \dim E(w_q) \cap E(x) \tag{7.7}$$

$$L_q = L_q(x,v,w) = E(x) \cap E(v_q) \cap E(w_{p-q}) \tag{7.8}$$

$$\dim L_q = 1 \tag{7.9}$$

$$E(x) = L_0 \oplus \cdots \oplus L_p \tag{7.10}$$

for $q = 0, 1, \cdots, p$. If $0 \neq \mathfrak{x}_q \in L_q$ for $q = 0, 1, \cdots, p$, then $x = P(\mathfrak{x}_0 \wedge \cdots \wedge \mathfrak{x}_p)$.

Assume $j \in \mathbf{Z}[0,p]$ exists such that $a_{j-1} > b_j^*$. Then $j > 0$. Define $\alpha = [13,8]$. By Proposition 8.15 of [1] the set R of all $(v,w) \in F(\mathfrak{a}) \times F(\mathfrak{b})$

such that a branch of $\alpha^{-1}(v,w)$ is contained in A is thin analytic in $F(\mathfrak{a}) \times F(\mathfrak{b})$. Take $(v,w) \in F(\mathfrak{a}) \times F(\mathfrak{b}) - R$. Define $\Phi = \alpha^{-1}(v,w) = \{(v,w)\} \times F(\mathfrak{c})$. Then $\Phi \cap A$ is thin analytic in Φ. Take $\tilde{z} \in F(\mathfrak{c})$ such that $(v,w,z) \in \Phi - A \subseteq N$. Therefore 7.6 holds and $M_q(v,w) = M_q = E(v_q) \cap E(w_{p-q})$ has dimension $a_q - b_q^* + 1 > 0$ for $q = 0, \cdots, p$. Also

$$\dim E(v_{j-1}) \cap E(w_{p-j}) \geq a_{j-1} + j + 1 + b_{p-j} + p - j - n - 1 = a_{j-1} - b_j^* > 0 .$$

Because $E(v_{j-1}) \cap E(w_{p-j})$ is contained in M_j and M_{j-1}, the intersection $M_j \cap M_{j-1}$ has positive dimension. Define $M = M_0 + \cdots + M_p$. Then

$$\dim M < \sum_{q=0}^{p} \dim M_q = \vec{\mathfrak{a}} + \vec{\mathfrak{b}} - d(p,n) + p + 1 = h + p + 1 .$$

Therefore $\dim M \leq h + p$. Hence $\hat{z}_0 \in G_{n-p-h}(V)$ exists such that $E(\hat{z}_0) \cap M = 0$. Take $\hat{z}_1, \cdots, \hat{z}_p$ such that $\hat{z} = (\hat{z}_0, \cdots, \hat{z}_p)$ belongs to $F(\mathfrak{c})$. An open neighborhood U of $\hat{z}$ in $F(\mathfrak{c})$ exists such that $E(z_0) \cap M = 0$ if $z = (z_0, \cdots, z_p) \in U$. Hence take $z = (z_0, \cdots, z_p) \in U$ such that $(v,w,z) \in \Phi - A \subseteq N$. Then $x \in G_p(V)$ exists such that $(x,v,w,z) \in G$. Now (7.7)-(7.10) hold, because $L_q \subseteq M_q$ for $q = 0, \cdots, p$, we have $E(x) \subseteq M$. Hence $E(x) \cap E(z_0) = 0$. However, $(x,z) \in S(\mathfrak{c})$ implies $\dim E(x) \cap E(z_0) \geq 1$. We have obtained a contradiction. Hence $a_{j-1} \leq b_j^* \leq a_j$ for $j = 0, 1, \cdots, p$.

We still have to prove that ζ is a modification. Hence $s = 1$ has to be shown. The set

$$A_3' = \bigcup_{0 \leq m < q \leq p} \{(v,w) \in F(\mathfrak{a}) \times F(\mathfrak{b}) \mid M_m(v,w) \cap M_q(v,w) \neq 0\}$$

is analytic. We will show that A_3' is thin. Let $e_0, \cdots, e_n$ be a base of V. For $q = 0, \cdots, p$ define

$$v_q = P(e_0 \wedge \cdots \wedge e_{\hat{a}_q}) \qquad w_q = P(e_n \wedge \cdots \wedge e_{n-\hat{b}_q}) .$$

Then $v = (v_0, \cdots, v_p) \in F(\mathfrak{a})$ and $w = (w_0, \cdots, w_p) \in F(\mathfrak{b})$. Also $E(w_{p-q})$ is spanned by $\mathfrak{e}_\mu$ for $\mu = \hat{b}^*_q, \cdots, n$. Take any $0 \leq m < q \leq p$. Then $M_m(v,w) \cap M_q(v,w)$ is contained in $E(v_m) \cap E(w_{p-m+1})$ where $E(v_m)$ is spanned by $\mathfrak{e}_\mu$ for $\mu = 0, \cdots, \hat{a}_m$ and $E(w_{p-m+1})$ is spanned by $\mathfrak{e}_\mu$ for $\mu = n, \cdots, \hat{b}^*_{m+1}$ with $\hat{a}_m = a_m + m \leq b^*_{m+1} + m = \hat{b}^*_{m+1} - 1 < \hat{b}^*_{m+1}$. Hence $E(v_m) \cap E(w_{p-m+1}) = 0$ and $M_m(v,w) \cap M_q(v,w) = 0$. We have $(v,w) \in F(\mathfrak{a}) \times F(\mathfrak{b}) - A'_3$. The analytic sets A'_3 and $A_3 = A'_3 \times F(\mathfrak{c})$ are thin in $F(\mathfrak{a}) \times F(\mathfrak{b})$ respectively in $F(\mathfrak{a}) \times F(\mathfrak{b}) \times F(\mathfrak{c})$. Therefore $N - A_3$ is open, connected and dense in $F(\mathfrak{a}) \times F(\mathfrak{b}) \times F(\mathfrak{c})$.

Take $(v,w,z') \in N - A_3$. Then $M_q = E(v_q) \cap E(w_{p-q})$ has positive dimension $a_q - b^*_q + 1$ and $M = M_0 + \cdots + M_p$ is a direct sum with

$$\dim M = \vec{\mathfrak{a}} + \vec{\mathfrak{b}} - d(p,n) + p + 1 = h + p + 1 .$$

Define $M^q = M_0 + \cdots + M_{q-1} + M_{q+1} + \cdots + M_p$ for $q = 0, \cdots, p$. Then $\dim M^q < h+p+1$. Also $z' = (z'_0, \cdots, z'_p)$ with $\dim E(z'_0) = n - h - p + 1$. Therefore arbitrary close to z', there exists a point $z = (z_0, \cdots, z_p)$ in $F(\mathfrak{c})$ such that $(v,w,z) \in N - A_3$ and such that

$$\dim M \cap E(z_0) = 1 \qquad \dim M^q \cap E(z_0) = 0$$

for $q = 0, 1, \cdots, p$.

Take (x,v,w,z) and (x',v,w,z) in $S(\mathfrak{a}, \mathfrak{b}, \mathfrak{c})$. Then (7.6)-(7.10) hold for x and for x'. Take $0 \neq x_q \in L_q(x,v,w)$ and $0 \neq x'_q \in L_q(x',v,w)$. Then $\mathfrak{x}_0, \cdots, \mathfrak{x}_p$ is a base of $E(x)$ and $\mathfrak{x}'_0, \cdots, \mathfrak{x}'_q$ is a base of $E(x')$. Because $L_q(x,v,w)$ and $L_q(x',v,w)$ are contained in M_q, we have $E(x) \subseteq M$ and $E(x') \subseteq M$. Take $0 \neq \mathfrak{z} \in M \cap E(z_0)$. Since $(x,z) \in F(\mathfrak{c})$, the linear subspace $E(x) \cap E(z_0)$ of $M \cap E(z_0)$ has at least dimension 1. Therefore $E(x) \cap E(z_0) = M \cap E(z_0)$. There exist unique complex numbers $\lambda_q \in \mathbb{C}$ such that

$$\mathfrak{z} = \sum_{q=0}^{p} \lambda_q \mathfrak{x}_q \in M .$$

Here $\mathfrak{x}_q \in M_q$ for $q = 0, \cdots, p$. Assume $\lambda_j = 0$ for some $j \in Z[0,p]$. Then $0 \neq z \in M^j \cap E(z_0) = 0$ which is impossible. Therefore $\lambda_q \neq 0$. By symmetry $0 \neq \lambda'_q \in C$ exists such that $\mathfrak{z} = \lambda'_0 \mathfrak{x}'_0 + \cdots + \lambda'_p \mathfrak{x}'_p$ with $\mathfrak{x}'_q \in M_q$ for $q = 0, \cdots, p$. Because M is the direct sum of $M_0, \cdots, M_p$, we have $\lambda_q \mathfrak{x}_q = \lambda'_q \mathfrak{x}'_q$ for $q = 0, \cdots, p$. Hence

$$x = P(\mathfrak{x}_0 \wedge \cdots \wedge \mathfrak{x}_p) = P(\mathfrak{x}'_0 \wedge \cdots \wedge \mathfrak{x}'_p) = x'.$$

The fiber $\zeta_0^{-1}(v,w,z)$ consists of at most one point where $(v,w,z) \in N$. Because the sheet number s of $\zeta_0 : G \to N$ is constant and positive, we have $s = 1$. Therefore ζ is a modification; q.e.d.

Parts of this proof are similar to the arguments of Chern [5], 367-369. However, Chern does not use the triple intersection diagram nor does he use fiber integration. He uses intersection theory. So the set up in [5] is different.

PROPOSITION 7.3. *Take* $\mathfrak{a}$ *and* $\mathfrak{b}$ *in* $\mathfrak{S}(p,n)$ and $h \in Z[0, n-p]$ *with* $\vec{\mathfrak{a}} + \vec{\mathfrak{b}} = d(p,n) + h$. *Assume* $c(\mathfrak{a}) \wedge c(\mathfrak{b}) \wedge c_h[p] \neq 0$. *Then* $a_{j-1} \leq b_j^* \leq a_j$ for $j = 0, 1, \cdots, p$ *and*

$$\int_{G_p(V)} c(\mathfrak{a}) \wedge c(\mathfrak{b}) \wedge c_h[p] = 1 \tag{7.11}$$

$$c(\mathfrak{a}) \wedge c(\mathfrak{b}) \wedge c(\mathfrak{c}) = c_{n-p}[p]^{p+1}.$$

Proof. Define $\mathfrak{c} = (n-p-h, n-p, \cdots, n-p) \in \mathfrak{S}(p,n)$. Then $c_h[p] = c(\mathfrak{c})$ by Proposition 5.4. Consider the triple intersection diagram 7.1. Define Ω by (7.2). Then (7.4) holds and implies $\eta_*(\zeta^*(\Omega)) \neq 0$. Therefore ζ is surjective. By Lemma 7.2, ζ is a modification and $a_{j-1} \leq b_j^* \leq a_j$ holds for $j = 0, \cdots, p$. Observe that $c(\mathfrak{a}) \wedge c(\mathfrak{b}) \wedge c_h[p]$ is a volume form on $G_p(V)$. Hence

$$\int_{G_p(V)} c(\mathfrak{a}) \wedge c(\mathfrak{b}) \wedge c_h[p]$$

$$= \int_{G_p(V)} \eta_* \zeta^*(\Omega) = \int_{S(\mathfrak{a},\mathfrak{b},\mathfrak{c})} \zeta^*(\Omega) =$$

$$= \int_{F(\mathfrak{a})\times F(\mathfrak{b})\times F(\mathfrak{c})} \Omega = \int_{F(\mathfrak{a})} \Omega_{\mathfrak{a}} \int_{F(\mathfrak{b})} \Omega_{\mathfrak{b}} \int_{F(\mathfrak{c})} \Omega_{\mathfrak{c}} = 1 .$$

Because $c_{n-p}[p]^{p+1} > 0$, a function γ exists such that

$$c(\mathfrak{a}) \wedge c(\mathfrak{b}) \wedge c_h[p] = \gamma c_{n-p}[p]^{p+1} .$$

Because $c(\mathfrak{a})$, $c(\mathfrak{b})$, $c_h[p]$ and $c_{n-p}[p]$ are invariant under the actions of the unitary group, γ is constant. Integration over $G_p(V)$ and Theorem 4.4 imply that $\gamma = 1$, q.e.d.

THEOREM 7.4 (Pieri [24]). *Take* $p \in \mathbb{Z}[0,n]$ *and* $h \in \mathbb{Z}[0, n-p]$. *Take* $\mathfrak{a} \in \mathfrak{S}(p,n)$. *Define* $a_{-1} = 0$. *Let* T *be the set of all* $\mathfrak{b} \in \mathfrak{S}(p,n)$ *with* $\vec{\mathfrak{b}} = \vec{\mathfrak{a}} - h$ *and* $a_{j-1} \leq b_j \leq a_j$ *for* $j = 0,\cdots,p$. *Then*

$$c[\mathfrak{a}] \wedge c_h[p] = \sum_{\mathfrak{b} \in T} c(\mathfrak{b}) .$$

Proof. Define $m = \vec{\mathfrak{a}} - h$. The form $c[\mathfrak{a}] \wedge c_h[p]$ has degree $2(d(p,n) - m)$. If $m < 0$, then $T = \emptyset$ and the theorem is true. Assume that $m \geq 0$. Then $\{c(\mathfrak{b})\}_{\mathfrak{b} \in \mathfrak{S}(p,n,m)}$ is a base of the vector space of all invariant forms of degree $2(d(p,n) - m)$; hence there exist numbers $\gamma_{\mathfrak{b}} \in \mathbb{C}$ such that

$$(7.12) \qquad c(\mathfrak{a}) \wedge c_h[p] = \sum_{\mathfrak{b} \in \mathfrak{S}(p,n,m)} \gamma_{\mathfrak{b}} c(\mathfrak{b}) .$$

Theorem 6.10 implies that $c(\mathfrak{a}) \wedge c(\mathfrak{b}^*) \wedge c_h[p] = \gamma_{\mathfrak{b}} c(\mathfrak{b}) \wedge c(\mathfrak{b}^*)$, where $c(\mathfrak{b}) \wedge c(\mathfrak{b}^*) > 0$. Here $\vec{\mathfrak{a}} + \vec{\mathfrak{b}}^* = d(p,n)+h$. If $\gamma_{\mathfrak{b}} \neq 0$, then $c(\mathfrak{a}) \wedge c(\mathfrak{b}^*) \wedge c_h[p] \neq 0$ and Proposition 7.3 implies that $\mathfrak{b} \in T$. Moreover Theorem 6.10, (7.11) and (7.12) imply that $\gamma_{\mathfrak{b}} = 1$, q.e.d.

For proofs of Pieri's theorem see also Hodge [17], Chern [5] and Vesentini [34].

c) *The Theorem of Giambelli*

The derivation of Giambelli's theorem from Pieri's theorem is standard and based on elementary computation. Hodge [18] gives the proof, whereas Chern [5] and Vesentini [34] state only the theorem. For completeness, the proof shall be included here.

THEOREM 7.5 (Giambelli [13]). *Take* $p \in Z[0,n]$ *and* $\mathfrak{a} \in \mathfrak{S}(p,n)$. *Define* $c_{jk} = c_{jk}(\mathfrak{a}) = c_{n-p-a_j-j+k}[p]$. *Then*

$$(7.13) \qquad c(\mathfrak{a}) = \begin{vmatrix} c_{00}, & \cdots, & c_{0p} \\ \vdots & & \vdots \\ c_{p0}, & \cdots, & c_{pp} \end{vmatrix}$$

Proof. Define $c[\mathfrak{a}] = \det c_{jk}$. Then $c[\mathfrak{a}] = c(\mathfrak{a})$ has to be proved. If $\mathfrak{a} = (n-p,\cdots,n-p)$, then $c_{jk} = 0$ if $j > k$ and $c_{jj} = 1$. Hence $c(n-p,\cdots,n-p) = 1 = c[n-p,\cdots,n-p]$ and (7.13) holds in this case.

The symbol $\mathfrak{a}$ is said to be of type q, if $q \in Z[0,p]$ and if $a_j = n-p$ for $j = q+1,\cdots,p$. If $\mathfrak{a}$ is of type q, then $c_{jk} = 0$ if $j > k$ and $j > q$. Also $c_{jj} = 1$ if $j > q$. Therefore

(7.14) $$c[\mathfrak{a}] = \begin{vmatrix} c_{00}, \cdots, c_{0q} \\ \cdot \quad\quad \cdot \\ \cdot \quad\quad \cdot \\ \cdot \quad\quad \cdot \\ \cdot \quad\quad \cdot \\ c_{q0}, \cdots, c_{qq} \end{vmatrix}$$

Now the theorem will be proved by induction for q. At first, consider $q = 0$. By Proposition 5.4, $c(\mathfrak{a}) = c(a_0, n-p, \cdots, n-p) = c_{n-p-a_0}[h]$. Also $c[\mathfrak{a}] = c_{00} = c_{n-p-a_0}[h]$. Hence $c(\mathfrak{a}) = c[\mathfrak{a}]$.

Assume the theorem has been proved for all symbols of type $q-1$, where $0 < q \leq p$. Then the theorem will be proved for symbols of type q. Take $\mathfrak{a} \in \mathfrak{S}(p,n)$ with $a_j = n-p$ for $j = q+1, \cdots, p$. For $\mu \in Z[0,q]$ define

$$\mathfrak{a}_\mu = (a_0-1, \cdots, a_{\mu-1}-1, a_{\mu+1}, \cdots, a_q, n-p, \cdots, n-p) .$$

Then $\mathfrak{a}_0 \in \mathfrak{S}(p,n)$ has type $q-1$. Hence $c(\mathfrak{a}_0) = c[\mathfrak{a}_0]$. Consider the case $1 \leq \mu \leq q$. If $a_0 > 0$, then $\mathfrak{a}_\mu \in \mathfrak{S}(p,n)$ has type $q-1$. Hence $c(\mathfrak{a}_\mu) = c[\mathfrak{a}_\mu]$. If $a_0 = 0$, then $\mathfrak{a} \notin \mathfrak{S}(p,n)$. Define $c(\mathfrak{a}_\mu) = 0$. Now, $c[\mathfrak{a}_\mu]$ is defined by the determinant with $c_{0k}(\mathfrak{a}_\mu) = c_{n-p+1+k}[p] = 0$ for $k = 0, \cdots, p$. Hence $c[\mathfrak{a}_\mu] = 0 = c(\mathfrak{a}_\mu)$. The development of the determinant gives

(7.15) $$\begin{aligned} c[\mathfrak{a}] &= \sum_{\mu=0}^{q} (-1)^\mu c_{n-p-\hat{a}_\mu}[p] \wedge c[\mathfrak{a}_\mu] \\ &= \sum_{\mu=0}^{q} (-1)^\mu c_{n-p-\hat{a}_\mu}[p] \wedge c(\mathfrak{a}_\mu) . \end{aligned}$$

Let T_0 be the set of all $\mathfrak{b} \in \mathfrak{S}(p,n)$ such that

$$\begin{aligned}
&\vec{\mathfrak{b}} = \vec{\mathfrak{a}} = \vec{\mathfrak{a}}_0 - (n-p-\hat{a}_0) \\
&0 \le b_0 \le a_1 \\
(7.16) \qquad &a_\lambda \le b_\lambda \le a_{\lambda+1} \qquad \text{for } \lambda = 1,\cdots,q-1 \\
&a_q \le b_q \le n-p \\
&b_\lambda = n-p = a_\lambda \qquad \text{for } \lambda = q+1,\cdots,p\,.
\end{aligned}$$

Now Theorem 7.4 implies that

$$(7.17) \qquad c_{n-p-\hat{a}_0}[p] \wedge c(\mathfrak{a}_0) = \sum_{\mathfrak{b}\in T_0} c(\mathfrak{b})\,.$$

If $1 \le \mu \le q-1$, let T_μ be the set of all $\mathfrak{b} \in \mathfrak{S}(p,n)$ such that

$$\begin{aligned}
&\vec{\mathfrak{b}} = \vec{\mathfrak{a}} = \vec{\mathfrak{a}}_\mu - (n-p-\hat{a}_\mu) \\
&0 \le b_0 < a_0 \\
&a_{\lambda-1}-1 \le b_\lambda < a_\lambda \qquad \text{for } \lambda = 1,\cdots,\mu-1 \\
&a_{\mu-1}-1 \le b_\mu \le a_{\mu+1} \\
&a_\lambda \le b_\lambda \le a_{\lambda+1} \qquad \text{for } \lambda = \mu+1,\cdots,q \\
&a_q \le b_q \le n-p \\
&b_\lambda = n-p = a_\lambda \qquad \text{for } \lambda = q+1,\cdots,p\,.
\end{aligned}$$

Let T_q be the set of all $\mathfrak{b} \in \mathfrak{S}(p,n)$ such that

$$\begin{aligned}
&\vec{\mathfrak{b}} = \vec{\mathfrak{a}} = \vec{\mathfrak{a}}_q - (n-p-\hat{a}_q) \\
&0 \le b_0 < a_0 \\
&a_{\lambda-1}-1 \le b_\lambda < a_\lambda \qquad \text{for } \lambda = 1,\cdots,q-1 \\
&a_{q-1}-1 \le b_q \le n-p \\
&b_\lambda = n-p = a_\lambda \qquad \text{for } \lambda = q+1,\cdots,p\,.
\end{aligned}$$

If $a_0 = 0$, then $T_\mu = \emptyset$. Hence

$$c_{n-p-\hat{a}_\mu}[p] \wedge c(\mathfrak{a}_\mu) = \sum_{\mathfrak{b} \in T_\mu} c(\mathfrak{b}) . \tag{7.18}$$

If $a_0 > 0$, then $\mathfrak{a}_\mu \in \mathfrak{S}(p,n)$ and Theorem 7.4 implies (7.18). Hence (7.15), (7.17) and (7.18) show that

$$c[\mathfrak{a}] = \sum_{\mu=0}^{q} (-1)^\mu \sum_{\mathfrak{b} \in T_\mu} c(\mathfrak{b}) . \tag{7.19}$$

For $\mu \in Z[0,q]$ let S_μ be the set of all $\mathfrak{b} \in \mathfrak{S}(p,n)$ such that

$$\begin{array}{ll} \vec{\mathfrak{b}} = \vec{\mathfrak{a}} & 0 \leq b_0 < a_0 \\ a_{\lambda-1} - 1 \leq b_\lambda < a_\lambda & \text{for } \lambda = 1,\cdots,\mu \\ a_\lambda \leq b_\lambda \leq a_{\lambda+1} & \text{for } \lambda = \mu+1,\cdots,q \\ b_\lambda = n-p = a_\lambda & \text{for } \lambda = q+1,\cdots,n . \end{array}$$

Also let S be the set of all $\mathfrak{b} \in \mathfrak{S}(p,n)$ such that

$$\begin{array}{ll} \vec{\mathfrak{b}} = \vec{\mathfrak{a}} & a_0 \leq b_0 \leq a_1 \\ a_\lambda \leq b_\lambda \leq a_{\lambda+1} & \text{for } \lambda = 1,\cdots,p . \end{array}$$

Then $T_0 = S_0 \cup S$ with $S_0 \cap S = \emptyset$, and $T_\mu = S_{\mu-1} \cup S_\mu$ with $S_{\mu-1} \cap S_\mu = \emptyset$ for $\mu = 1,\cdots,q$. Define

$$A = \sum_{\mathfrak{b} \in S} c(\mathfrak{b}) \qquad A_\mu = \sum_{\mathfrak{b} \in S_\mu} c(\mathfrak{b})$$

for $\mu = 0,\cdots,q$. Thus (7.19) implies

$$c[\mathfrak{a}] = A + A_0 + \sum_{\mu=1}^{q} (-1)^\mu (A_{\mu-1} + A_\mu) = A + (-1)^q A_q .$$

Clearly $\mathfrak{a} \in S$. If $\mathfrak{b} \in S$, then $a_\lambda \leq b_\lambda$ for $\lambda = 0,\cdots,p$. If $a_j < b_j$ for some j, then $\vec{\mathfrak{a}} < \vec{\mathfrak{b}}$ which is wrong. Hence $\mathfrak{b} = \mathfrak{a}$ and $S = \{\mathfrak{a}\}$. Therefore $A = c(\mathfrak{a})$. If $\mathfrak{b} \in S_q$, then $b_\lambda < a_\lambda$ for $\lambda = 0,\cdots,q$. Hence $\vec{\mathfrak{b}} < \vec{\mathfrak{a}}$, which is impossible. Therefore $S_q = \emptyset$ and $A_q = 0$. Consequently $c[\mathfrak{a}] = A = c(\mathfrak{a})$; q.e.d.

The invariant forms on $G_p(V)$ form a graded exterior algebra

$$\mathrm{Inv}(G_p(V)) = \bigoplus_{m=0}^{d(p,n)} \mathrm{Inv}^{2m}(G_p(V))$$

which is isomorphic to the cohomology ring of $G_p(V)$. Giambelli's theorem shows that $\mathrm{Inv}(G_p(V))$ is generated by the basic Chern forms $c_1[p],\cdots,c_{n-p}[p]$. Thus any invariant form can be expressed as a linear combination over $\mathbb{C}$ of wedge products of basic Chern forms. Each invariant form of degree $2m$ is a linear combination over $\mathbb{C}$ of monomials

(7.20) $$c_1[p]^{j_1} \wedge c_2[p]^{j_2} \wedge \cdots \wedge c_{n-p}[p]^{j_{n-p}} .$$

In fact, if we inspect the determinant (7.13) we see that the monomials with only $p+1$ factors suffice, i.e., each invariant form of degree $2m$ is a linear combination over $\mathbb{C}$ of monomials (7.20) with the restrictions

(7.21) $$0 \leq j_\lambda \in \mathbb{Z} \qquad \text{for } \lambda = 1,\cdots,n-p$$

(7.22) $$j_1 + j_2 + \cdots + j_{n-p} \leq p+1$$

(7.23) $$j_1 + 2j_2 + \cdots + (n-p)j_{n-p} = m .$$

Let $\mathfrak{H}_n(p,m)$ be the set of all $(j_1,\cdots,j_{n-p})$ satisfying the conditions (7.21), (7.22) and (7.23).

THEOREM 7.6. *Take* $p \in \mathbb{Z}[0,n)$ *and let* m *be an integer with* $0 \leq m \leq d(p,n)$. *Then the map which assigns to each* $(j_1,\cdots,j_{n-p}) \in \mathfrak{H}_n(p,m)$ *the monomial* (7.20) *is bijective and the set*

$$\{c_1[p]^{j_1} \wedge \cdots \wedge c_{n-p}[p]^{j_{n-p}} \mid (j_1, \cdots, j_{n-p}) \in \mathfrak{H}_n(p,m)\}$$

is a base over $\mathbb{C}$ *of the vector space* $\mathrm{Inv}^{2m}(G_p(V))$ *of all invariant forms of degree* $2m$ *on* $G_p(V)$.

Proof. We have seen that the set generates $\mathrm{Inv}^{2m}(G_p(V))$. Hence it suffices to show that $\# \mathfrak{H}_n(p,m) = \dim \mathrm{Inv}^{2m}(G_p(V)) = \# \mathfrak{S}(p,n,m)$ (see (2.12)). A bijective map

$$\phi : \mathfrak{H}_n(p,m) \to \mathfrak{S}(p,n,m)$$

shall be defined. Take $j = (j_1, \cdots, j_{n-p}) \in \mathfrak{H}_n(p,m)$. Define

$$r_\lambda = j_\lambda + j_{\lambda+1} + \cdots + j_{n-p} .$$

Then

$$p+1 \geq r_1 \geq r_2 \geq \cdots \geq r_{n-p} \geq 0 .$$

For $q = 0, 1, \cdots, p+1$, define

$$A_q = \{\lambda \in \mathbb{N}[1, n-p] \mid r_\lambda \geq p+1-q\} .$$

Then $A_0 \leq A_1 \subseteq \cdots \subseteq A_{p+1} = \mathbb{N}[1, n-p]$. Define $a_q = \# A_q$. Then $a_0 \leq a_1 \leq \cdots \leq a_p \leq a_{p+1} = n-p$. Hence $\mathfrak{a} = (a_0, \cdots, a_p) \in \mathfrak{S}(p,n)$. Define $A_{-1} = \emptyset$ and

$$B_q = A_q - A_{q-1} = \{\lambda \in \mathbb{N}[1, n-p] \mid r_\lambda = p+1-q\} .$$

Then $\alpha_q = \# B_q$ and $a_q = \alpha_0 + \cdots + \alpha_q$. Hence

$$m = \sum_{\nu=1}^{n-p} \nu j_\nu = \sum_{\nu=1}^{n-p} \sum_{\lambda=1}^{\nu} j_\nu = \sum_{\lambda=1}^{n-p} \sum_{\nu=\lambda}^{n-p} j_\nu = \sum_{\lambda=1}^{n-p} r_\lambda = \sum_{q=0}^{p+1} \alpha_q(p+1-q)$$

(7.24)

$$= \sum_{q=0}^{p} \sum_{s=q}^{p} \alpha_q = \sum_{s=0}^{p} \sum_{q=0}^{s} \alpha_q = \sum_{s=0}^{p} a_s = \vec{\mathfrak{a}} \, .$$

Therefore $\mathfrak{a} \in \mathfrak{S}(p,n,m)$. Define $\phi(j) = \mathfrak{a}$. The map ϕ is defined. Because r_λ is decreasing in λ we have

(7.25) $$B_q = \{\lambda \in N \mid a_{q-1} < \lambda \leq a_q\} ,$$

where $a_{-1} = 0$. If j and j' are given in $\mathfrak{H}_n(p,m)$ with $\phi(j) = \phi(j') = \mathfrak{a}$, then r_λ, r'_λ, A_q, A'_q and B_q, B'_q are defined. Obviously $B_q = B'_q$ and $r_\lambda = p+1-q = r'_\lambda$ for all $\lambda \in B_q$ and $q = 0,\cdots,p+1$. Because $B_0 \cup \cdots \cup B_{p+1} = N[1,n-p]$, we have $r_\lambda = r'_\lambda$ for all $\lambda = 1,\cdots,n-p$. Hence $j = j'$, the map ϕ is injective. Take $\mathfrak{a} = (a_0,\cdots,a_p) \in \mathfrak{S}(p,n,m)$. Define $a_{-1} = 0$ and $a_{p+1} = n-p$. Then B_q is defined by (7.25) and $N[1,n-p] = B_0 \cup \cdots \cup B_{p+1}$ is a disjoint union. Hence a decreasing function r_λ for $\lambda = 1,\cdots,n-p$ is defined by $r_\lambda = p+1-q$ for $\lambda \in B_q$. Define $j_\lambda = r_{\lambda+1} - r_\lambda \geq 0$ for $\lambda = 1,\cdots,n-p-1$ and $j_p = r_{n-p} \geq 0$. Then $j_1 + \cdots + j_{n-p} = r_1 \leq p+1$. The computation (7.24) shows $j_1 + 2j_2 + \cdots + (n-p)\,j_{n-p} = m$. Hence $j = (j_1,\cdots,j_{n-p}) \in \mathfrak{H}_n(p,m)$. Obviously, $\phi(j) = \mathfrak{a}$. Therefore ϕ is bijective. Hence $\#\,\mathfrak{H}_n(p,m) = \#\,\mathfrak{S}(p,n,m) = \dim \mathrm{Inv}^{2m}(G_p(V))$; q.e.d.

In many cases, p is kept fixed and $n \to \infty$. Then the infinite dimensional Grassmann manifold of p-planes is to be considered. The base constructed in Theorem 7.6 is not convenient for this purpose, since the length of the monomials depends on the codimension $n-p$. A base given by the Chern forms $s_0[p],\cdots,s_{p+1}[p]$ of $S_p(V)$ would be more convenient, since the number of generators depends on p only. Such a base shall be constructed now. Using (3.22) would be rather cumbersome. Duality shall be applied instead. Some preparations are needed.

If $\mathfrak{x} \in \bigwedge_p V$ and $\alpha \in \bigwedge_q V^*$ with $0 \leq q \leq p$ define $\mathfrak{x} \llcorner \alpha = \alpha(\mathfrak{x}) \in \mathbb{C}$ if $p = q$ and $\mathfrak{x} \llcorner \alpha \in \bigwedge_{p-q} V$ by $(\mathfrak{x} \llcorner \alpha) \wedge \beta = \mathfrak{x} \llcorner \alpha \wedge \beta$ for all $\beta \in \bigwedge_{p-q} V^*$ if $q < p$. Since $V = (V^*)^*$, also $\alpha \llcorner \mathfrak{x} \in \bigwedge_{q-p} V^*$ is defined if $\mathfrak{x} \in \bigwedge_p V$ and $\alpha \in \bigwedge_q V^*$ with $p \leq q$. If $p = q$, then $\alpha \llcorner \mathfrak{x} = \mathfrak{x} \llcorner \alpha$. If $a \in G_q(V^*)$, take $0 \neq \alpha = \alpha_0 \wedge \cdots \wedge \alpha_q \in \widetilde{G}_q(V^*)$. Then

$$E[a] = \{ \mathfrak{x} \in V \mid \alpha \llcorner \mathfrak{x} = 0 \} = \bigcap_{j=0}^{q} \ker \alpha_j$$

is a linear subspace of dimension $n-q$ of V which does not depend on the choices of $\alpha, \alpha_0, \cdots, \alpha_q$ to represent a. If $p \in \mathbb{Z}[0,n)$ and $q = n-p-1$, a biholomorphic map

$$\delta : G_p(V) \to G_q(V^*)$$

is defined by $E[\delta(x)] = E(x)$. If $0 \neq \mathfrak{e} \in \widetilde{G}_n(V)$, then a linear isomorphism $D_{\mathfrak{e}} : \bigwedge_{p+1} V \to \bigwedge_{q+1} V^*$ is defined by $(\mathfrak{y} \llcorner D_{\mathfrak{e}}\, \mathfrak{x})\, \mathfrak{e} = \mathfrak{y} \wedge \mathfrak{x}$ for all $\mathfrak{y} \in \bigwedge_{q+1} V$ and $\mathfrak{x} \in \bigwedge_{p+1} V$ such that $D_{\mathfrak{e}}(\widetilde{G}_p(V)) = \widetilde{G}_q(V^*)$. Then $\mathbb{P} \circ D_{\mathfrak{e}} = \delta \circ \mathbb{P}$ on $G_p(V) - \{0\}$. This shows that δ is biholomorphic. The map δ is called the *dualism*. Also a dualism can be constructed for V^*. Then $\delta = \delta^{-1}$.

The classifying sequence

$$0 \longrightarrow S_p(V) \xrightarrow[j]{} G_p(V) \times V \xrightarrow[\eta]{} Q_p(V) \longrightarrow 0 \tag{7.26}$$

defines a dual exact sequence

$$0 \longrightarrow Q_p(V)^* \xrightarrow[\eta^*]{} G_p(V^*) \times V^* \xrightarrow[j^*]{} S_p(V)^* \longrightarrow 0 \tag{7.27}$$

which also exists for the dual vector space V^*

$$0 \longrightarrow Q_q(V^*)^* \xrightarrow[\eta^*]{} G_q(V^*) \times V \xrightarrow[j^*]{} S_q(V^*)^* \longrightarrow 0\,. \tag{7.28}$$

LEMMA 7.7. *If* η^* *in* (7.27) *and* (7.28) *is identified with the inclusion map, then*

$$Q_p(V)^* = \{(x,\alpha) \in G_p(V) \times V^* \mid \alpha | E(x) = 0\}$$

$$Q_q(V^*)^* = \{(a,\mathfrak{x}) \in G_q(V^*) \times V \mid \mathfrak{x} \in E[a]\}$$

Proof. Take $x \in G_p(V)$. Pick $\alpha \in (Q_p(V)^*)_x$. Then $\alpha : Q_p(V)_x \to \mathbf{C}$ is linear. Hence $\alpha \circ \eta_x : V \to \mathbf{C}$ is defined and linear. Observe $\eta^*(\alpha) = \alpha \circ \eta_x$. Since $E(x)$ is the kernel of η_x, we have $\eta^*(\alpha) | E(x) = 0$. If η^* is identified with the inclusion, we obtain $\alpha | E(x) = 0$ where $\alpha \in V^*$. Take $\alpha \in V^*$ with $\alpha | E(x) = 0$. Then there exists a linear map $\beta : V/E(x) \to \mathbf{C}$ such that $\alpha = \beta \circ \eta_x$. Here $Q_p(V)_x = V/E(x)$ and $\eta_x : V \to V/E(x)$ is the residual map. Hence $\beta \in (Q_p(V)_x)^*$ with $\eta^*(\beta) = \alpha$. If η^* is regarded as an inclusion map, then $\alpha = \beta \in (Q_p(V)^*)_x$. The first identity is proved.

Take $a \in G_p(V^*)$. Then $a = \mathbf{P}(\alpha_0 \wedge \cdots \wedge \alpha_p)$ with $\alpha_\mu \in V^*$. Moreover $E(a) = \mathbf{C}\alpha_0 + \cdots + \mathbf{C}\alpha_p$. Take $\mathfrak{x} \in V$. Then $\mathfrak{x} | E(a) = 0$ if and only if $\mathfrak{x}(\alpha_\mu) = \alpha_\mu(\mathfrak{x}) = 0$ for $\mu = 0, \cdots, p$, which is the case if and only if $\mathfrak{x} \in E[a]$. The second identity is proved; q.e.d.

LEMMA 7.8. *If* $p \in \mathbf{Z}[0,n)$ *and* $q = n-p-1$, *then the exact sequence* (7.28) *pulls back to the exact sequence* (7.26) *under the dual map* $\delta : G_p(V) \to G_q(V^*)$. *Moreover, the hermitian metrics along the fibers of* (7.28) *pull back to the hermitian metrics given in* (7.26). *Hence*

$$s_\mu[p,V] = \delta^*(c_\mu[q,V^*]) \qquad \text{for } \mu = 0, \cdots, p+1 \tag{7.29}$$

$$c_\nu[p,V] = \delta^*(s_\nu[q,V^*]) \qquad \text{for } \nu = 0, \cdots, n-p. \tag{7.30}$$

Proof. The standard pull back of $Q_q(V^*)^*$ by δ is given by

$$\delta^* Q_q(V^*)^* = \{(x, \delta(x), \mathfrak{z}) \in G_p(V) \times G_q(V^*) \times V \mid \mathfrak{z} \in E[\delta(x)]\}.$$

Identify $(x, \delta(x), \mathfrak{z})$ with $(x, \mathfrak{z})$. Then

$$\delta^* Q_q(V^*)^* = \{(x, \mathfrak{z}) \in G_p(V) \times V \mid \mathfrak{z} \in E[\delta(x)]\}$$
$$= \{(x, \mathfrak{z}) \in G_p(V) \times V \mid \mathfrak{z} \in E(x)\} = S_p(V) .$$

Trivially $G_q(V^*) \times V$ pulls back to $G_p(V) \times V$. Hence $S_q(V^*)^*$ pulls back to $Q_q(V)$. By the constructions of the hermitian metrics along the fibers in (7.28) and (7.26), the hermitian metrics in (7.28) pull back to the hermitian metrics in (7.26) which implies (7.29) and (7.30).

THEOREM 7.9. *Take* $p \in \mathbb{Z}[0,n)$ *and let* m *be any integer with* $0 \leq m \leq d(p,n)$. *Let* $\mathfrak{B}_n(p,m)$ *be the set of all* $(j_1, \cdots, j_{p+1})$ *where* $j_\lambda \geq 0$ *are integers for* $\lambda = 0, \cdots, p$ *such that* $j_1 + \cdots + j_{p+1} \leq n-p$ *and such that*

$$j_1 + 2j_2 + 3j_3 + \cdots + (p+1) j_{p+1} = m .$$

Then the set

$$\{s_1[p]^{j_1} \wedge \cdots \wedge s_{p+1}[p]^{j_{p+1}} \mid (j_1, \cdots, j_{p+1}) \in \mathfrak{B}_n(p,m)\}$$

is a base over $\mathbb{C}$ *of the vector space* $\mathrm{Inv}^{2m}(G_p(V))$ *of all invariant forms of degree* $2m$ *on* $G_p(V)$, *and the base set is indexed bijectively.*

Proof. Define $q = n-p-1$. Then $n-p = q+1$ and $p+1 = n-q$. The duality map $\delta : G_p(V) \to G_q(V^*)$ is defined and the pull back

$$\delta^* : \mathrm{Inv}^{2m}(G_q(V^*)) \to \mathrm{Inv}^{2m}(G_p(V))$$

is an isomorphism. Because of (7.29) and (7.30) the base of $\mathrm{Inv}^{2m}(G_q(V^*))$ as given by Theorem 7.6 and (7.21), (7.22) and (7.23) pulls back to a base of $\mathrm{Inv}^{2m}(G_p(V))$ as described in Theorem 7.9, q.e.d.

If $n \to \infty$, the restriction $j_1 + \cdots + j_{p+1} \leq n-p$ can be lifted and the cohomology ring of the infinite dimensional Grassmann-manifold of order p is isomorphic to the free exterior algebra generated by $1, s_1[p], \cdots, s_{p+1}[p]$

over $\mathbf{C}$. From this the cohomology ring of the $G_p(V)$ with $\dim V = n+1 < \infty$ is obtained by truncating the ring by the rule $j_1 + \cdots + j_{p+1} \leq n-p$. Of course all this is well known, for instance see Chern [4] or Milnor-Stasheff [23], where the coefficient ring is $\mathbf{Z}$.

APPENDIX

The Schubert variety is an irreducible analytic set and the open Schubert cell is biholomorphically equivalent to an Euclidean space. This is well known. A proof can be found in Chern [7]. For the convenience of the reader, a proof is given here following the lines of Chern's proof.

THEOREM. *Let* V *be a complex vector space of dimension* $n+1$. *Take* $p \in \mathbf{Z}[0,n]$ *and* $\mathfrak{a} \in \mathfrak{S}(p,n)$. *Let* $v \in \mathbf{F}^n$ *be a complete flag. Then* $S\langle v,\mathfrak{a}\rangle$ *is an* $\vec{\mathfrak{a}}$*-dimensional irreducible analytic subset of* $G_p(V)$. *The open Schubert cell* $S^*\langle v,\mathfrak{a}\rangle$ *is biholomorphically equivalent to* $\mathbf{C}^{\vec{\mathfrak{a}}}$ *and is dense in* $S\langle v,\mathfrak{a}\rangle$.

Proof. If $p=0$, then $\mathfrak{a} = a_0 = a \in \mathbf{Z}[0,n]$ and $S\langle v,\mathfrak{a}\rangle = \ddot{E}(v_a)$. If $a=0$, then $\ddot{E}(v_0) = \{v_0\} = S^*\langle v,\mathfrak{a}\rangle$. If $a>0$, then $\partial_0 S\langle v,\mathfrak{a}\rangle = \ddot{E}(v_{a-1})$ and $S^*\langle v,\mathfrak{a}\rangle = \ddot{E}(v_a) - \ddot{E}(v_{a-1})$ is dense in $S\langle v,\mathfrak{a}\rangle$ and is biholomorphically equivalent to $\mathbf{C}^a$. The theorem holds in the case $p=0$.

Now, the theorem shall be proved for $p \geq 1$ under the assumption that it holds for $p-1$. If $\vec{\mathfrak{a}} = 0$, then $S\langle v,\mathfrak{a}\rangle = \{v_q\} = S^*\langle v,\mathfrak{a}\rangle$ and the theorem is trivially true. Assume $\vec{\mathfrak{a}} > 0$ and assume the theorem already has been proved for all $\mathfrak{b} \in \mathfrak{S}(p,n)$ with $\vec{\mathfrak{b}} < \vec{\mathfrak{a}}$.

Abbreviate $a = \vec{\mathfrak{a}}$ and $b_\mu = a_\mu = a_\mu + \mu$ for $\mu = 0,\cdots,p$. Observe that the flag $v = (v_0,\cdots,v_n) \in \mathbf{F}^n$ is given. Abbreviate

$$\text{(A1)} \qquad E(\mu) = E(v_\mu) \qquad \ddot{E}(\mu) = \ddot{E}(v_\mu)$$

for $\mu = 0,\cdots,n$. Define $Q = S\langle v,\mathfrak{a}\rangle - \partial_0 S\langle v,\mathfrak{a}\rangle$. Then Q is open in $S\langle v,\mathfrak{a}\rangle$. If $a_0 = 0$, then $\partial_0 S\langle v,\mathfrak{a}\rangle = 0$ and $Q = S\langle v,\mathfrak{a}\rangle$. If $a_0 > 0$, then

(A2) $$Q = \{x \in S<v, \mathfrak{a}> \mid \dim E(x) \cap E(b_0-1) = 0\}$$

CLAIM 1. Q is dense in $S<v,\mathfrak{a}>$.

Proof of Claim 1. Obviously, $a_0 > 0$ can be assumed. Take $x \in S^*<v, \partial_0 \mathfrak{a}>$. Lemma 2.2 implies that

$$\dim E(x) \cap E(b_\mu - 1) = \mu \qquad \dim E(x) \cap E(b_\mu) = \mu + 1$$

for $\mu = 1, \cdots, p$ and $\dim E(x) \cap E(b_0 - 1) = 1$. Therefore a basis $\mathfrak{r}_0, \cdots, \mathfrak{r}_n$ of V and $\mathfrak{q} \in V$ exist such that

$$E(b_0-1) = \mathbb{C}\mathfrak{r}_0 + \cdots + \mathbb{C}\mathfrak{r}_{b_0-1}$$

$$E(b_\mu) = \mathbb{C}\mathfrak{r}_0 + \cdots + \mathbb{C}\mathfrak{r}_{b_\mu} \qquad \text{for } \mu = 0, 1, \cdots, p$$

$$E(x) \cap E(b_0-1) = \mathbb{C}\mathfrak{r}_{b_0-1}$$

$$E(x) = \mathbb{C}\mathfrak{r}_{b_0-1} + \mathbb{C}\mathfrak{q} + \mathbb{C}\mathfrak{r}_{b_2} + \cdots + \mathbb{C}\mathfrak{r}_{b_p}$$

$$0 \neq \mathfrak{q} \in \mathbb{C}\mathfrak{r}_{b_0+1} + \cdots + \mathbb{C}\mathfrak{r}_{b_1} .$$

If $0 \neq \lambda \in \mathbb{C}$, then

$$\mathfrak{x}_\lambda = (\mathfrak{r}_{b_0-1} + \lambda \mathfrak{r}_{b_0}) \wedge \mathfrak{q} \wedge \mathfrak{r}_{b_2} \wedge \cdots \wedge \mathfrak{r}_{b_p} \neq 0 .$$

Define $x_\lambda = P(\mathfrak{x}_\lambda) \in G_p(V)$. Then $x_\lambda \to x$ for $\lambda \to 0$. If $1 \leq \mu \leq p$, then

$$E(x_\lambda) \cap E(b_\mu) \supseteq \mathbb{C}(\mathfrak{r}_{b_0-1} + \lambda \mathfrak{r}_{b_0}) + \mathbb{C}\mathfrak{q} + \mathbb{C}\mathfrak{r}_{b_2} + \cdots + \mathbb{C}\mathfrak{r}_{b_\mu} .$$

Hence $\dim E(x_\lambda) \cap E(b_\mu) \geq \mu + 1$. Also $\mathfrak{r}_{b_0-1} + \lambda \mathfrak{r}_{b_0} \in E(x_\lambda) \cap E(b_0)$. Hence $\dim E(x_\lambda) \cap E(b_0) \geq 1$. Therefore $x_\lambda \in S<v, \mathfrak{a}>$. Also $E(x_\lambda) \cap E(b_0-1) = 0$. Hence $x_\lambda \in Q$ by (A2). Now, $\lambda \to 0$ implies that $x \in \overline{Q}$. Therefore $S^*<v, \partial_0 \mathfrak{a}> \subseteq \overline{Q}$. By induction assumption $S^*<v, \partial_0 \mathfrak{a}>$ is dense in $S<v, \partial_0 \mathfrak{a}>$. Hence $\partial_0 S<v, \mathfrak{a}> \subseteq \overline{Q}$. Claim 1 is proved.

Take a base $\mathfrak{e}_0, \cdots, \mathfrak{e}_n$ of V such that $v_\mu = P(\mathfrak{e}_0 \wedge \cdots \wedge \mathfrak{e}_\mu)$ for

$\mu = 0,\cdots,n$. Define $w_\mu = v_\mu$ for $\mu = 0,1,\cdots,a_0-1$ and

(A3) $$w_\mu = \mathbf{P}(\mathfrak{e}_0 \wedge \cdots \wedge \mathfrak{e}_{a_0-1} \wedge \mathfrak{e}_{a_0+1} \wedge \cdots \wedge \mathfrak{e}_{\mu+1})$$

for $\mu = a_0,\cdots,n-1$. Define $W = E(w_{n-1})$ and $L = \mathbb{C}\,\mathfrak{e}_{a_0}$. Then $V = W \oplus L$. Define

(A4) $$A = \ddot{E}(a_0) - \ddot{E}(a_0-1)$$

(A5) $$B = \{y \in G_{p-1}(W) \mid E(y) \cap E(a_0-1) = 0\} .$$

Then B is open in $G_{p-1}(V)$ and $G_{p-1}(V) - B$ is analytic. Here A is biholomorphically equivalent to $\mathbb{C}^{a_0}$.

A holomorphic map $f: Q \to A \times B$ shall be defined. By Lemma 1.4

$$\Gamma = \{(x,y) \in S\langle v,\mathfrak{a}\rangle \times \ddot{E}(a_0) \mid y \in \ddot{E}(x)\}$$

is analytic. By the definition of $S\langle v,\mathfrak{a}\rangle$, the projection $\pi: \Gamma \to S\langle v,\mathfrak{a}\rangle$ is surjective. Take $x \in Q$. Then $\dim E(x) \cap E(a_0) \geq 1$ and $\dim E(x) \cap E(a_0-1) = 0$. Hence $\dim E(x) \cap E(a_0) = 1$. Therefore $\ddot{E}(x) \cap \ddot{E}(a_0) = \{g(x)\}$ consists of exactly one point. Observe $g(x) \notin \ddot{E}(a_0-1)$. Therefore a map $g: Q \to A$ is defined, whose graph $\Gamma \cap (Q \times \ddot{E}(a_0))$ is analytic. Hence g is holomorphic.

Take $x \in Q$. Then $g(x) = \mathbf{P}(\mathfrak{g})$ with $0 \neq \mathfrak{g} \in E(x) \cap (E(a_0) - E(a_0-1))$. Since $W \cap E(a_0) = E(a_0-1)$, we conclude that $\mathfrak{g} \notin W$. Hence $\dim W \cap E(x) = p$. Therefore one and only one $h(x) \in G_{p-1}(W)$ exists such that

(A6) $$E(h(x)) = E(x) \cap W .$$

Because $E(h(x)) \cap E(a_0-1) = E(x) \cap W \cap E(a_0-1) = 0$, we have $h(x) \in B$. A map $h: Q \to B$ is defined. We will show that h is holomorphic.

Take $x_0 \in Q$. An open neighborhood U of x_0 in Q and a holomorphic vector function $\mathfrak{g}: U \to V$ exists such that $g(x) = \mathbf{P}(\mathfrak{g}(x))$ for all $x \in U$. Then $\mathfrak{g}(x) \in E(x) \cap (E(a_0) - E(a_0-1)$. Holomorphic maps $\mathfrak{g}_0: U \to E(a_0-1)$ and $g_0: U \to \mathbb{C} - \{0\}$ exist such that $\mathfrak{g}(x) = \mathfrak{g}_0(x) + g_0(x)\,\mathfrak{e}_{a_0}$

for all $x \in U$. Since the tautological bundle is locally trivial, U can be taken so small that there exist holomorphic vector functions $\mathfrak{y}_\mu : U \to V$ for $\mu = 1, \cdots, p$ such that $\mathfrak{g}, \mathfrak{y}_1, \cdots, \mathfrak{y}_p$ is a holomorphic frame of $S_p(V)$ over U; i.e. $\mathfrak{g}(x), \mathfrak{y}_1(x), \cdots, \mathfrak{y}_p(x)$ is a base of $E(x)$ for each $x \in U$. Holomorphic maps $\mathfrak{y}_\mu^0 : U \to W$ and $y_\mu : U \to \mathbb{C}$ exist such that $\mathfrak{y}_\mu = \mathfrak{y}_\mu^0 + y_\mu \, e_{a_0}$ on U. Because $E(a_0 - 1) \subseteq W$, we have $\mathfrak{g}_0(x) \in W$ if $x \in U$. Then

$$\mathfrak{z}_\mu = \mathfrak{y}_\mu - \frac{y_\mu}{g_0} \mathfrak{g} = \mathfrak{y}_\mu^0 - \frac{y_\mu}{g_0} \mathfrak{g}_0 : U \to W$$

is holomorphic. For each $x \in U$, the vectors $\mathfrak{z}_1(x), \cdots, \mathfrak{z}_p(x)$ define a base of $E(x) \cap W = E(h(x))$. Hence $h(x) = \mathbf{P}(\mathfrak{z}_1(x) \wedge \cdots \wedge \mathfrak{z}_p(x))$ if $x \in U$ and $h|U$ is holomorphic. Therefore h is holomorphic on Q. A holomorphic map

$$\text{(A7)} \qquad f = (g,h) : Q \to A \times B$$

is defined.

CLAIM 2. The map f is injective.

Proof of Claim 2. Take x and $\tilde{x}$ in Q with $f(x) = f(\tilde{x})$. Then $g(x) = g(\tilde{x}) = \mathbf{P}(\mathfrak{g})$ with $\mathfrak{g} \notin W$. Also $h(x) = h(\tilde{x})$. Therefore

$$E(x) = (E(x) \cap W) \oplus \mathbb{C}\,\mathfrak{g} = E(h(x)) \oplus \mathbb{C}\,\mathfrak{g} =$$
$$= E(h(\tilde{x})) \oplus \mathbb{C}\,\mathfrak{g} = (E(\tilde{x}) \cap W) \oplus \mathbb{C}\,\mathfrak{g} = E(\tilde{x}) .$$

Hence $x = \tilde{x}$; the second claim is proved.

Recall the definition of w_μ in (A3). Then $w = (w_0, \cdots, w_{n-1}) \in F^{n-1}$ is a total flag for W. Define $c_\mu = a_{\mu+1}$ and $d_\mu = c_\mu + \mu = \hat{c}_\mu$ for $\mu = 0, 1, \cdots, p-1$. Then $\mathfrak{c} = (c_0, \cdots, c_{p-1}) \in \mathfrak{S}(p-1, n-1)$ is a symbol and the Schubert variety $S\langle w, \mathfrak{c}\rangle$ in $G_{p-1}(W)$ is defined. Here $W \subseteq V$ implies that $G_{p-1}(W) \subseteq G_{p-1}(V)$. Define $E'(\mu) = E(w_\mu)$ for $\mu = 0, 1, \cdots, n-1$. Then

$$\text{(A8)} \qquad E(\mu) \cap W = E'(\mu-1) \qquad \text{if } a_0 \le \mu \le n$$

$$\text{(A9)} \qquad E(\mu) \cap W = E(\mu) = E'(\mu) \qquad \text{if } 0 \le \mu < a_0 \,.$$

Define $C = B \cap S\langle w, \mathfrak{c}\rangle$.

CLAIM 3. If $x \in Q$, then $h(x) \in C$.

Proof of Claim 3. Take $\mu \in \mathbf{Z}[0, p-1]$. Then $b_{\mu+1} \ge a_0$. Hence (A8) implies that $W \cap E(b_{\mu+1}) = E'(b_{\mu+1}-1) = E'(d_\mu)$. Therefore

$$\begin{aligned} \dim E(h(x)) \cap E'(d_\mu) &= \dim E(x) \cap W \cap E(b_{\mu+1}) \\ &\ge \dim E(x) \cap E(b_{\mu+1}) + \dim W - \dim V \ge \mu + 1 \,. \end{aligned}$$

Hence $h(x) \in S\langle w, \mathfrak{c}\rangle$. Also $h(x) \in B$. Claim 3 is proved.

CLAIM 4. The map $f: Q \to A \times C$ is bijective.

Proof of Claim 4. Take $(y,z) \in A \times C$. Then $E(z) \subseteq W$ and $y = \mathbf{P}(\mathfrak{y})$ with $0 \neq \mathfrak{y} \in E(a_0) - E(a_0-1)$. Since $W \cap E(a_0) = E(a_0-1)$, this implies that $\mathfrak{y} \notin W$. Therefore $\mathfrak{y} \notin E(z)$. One and only one $x \in G_p(V)$ exists such that $E(x) = \mathbf{C}\mathfrak{y} + E(z)$. Then $\mathfrak{y} \in E(x) \cap E(b_0)$. Hence $\dim E(x) \cap E(b_0) \ge 1$. If $1 \le \mu \le p$, then $b_\mu \ge a_0$. Now (A8) implies that $E(b_\mu) \cap W = E'(b_\mu-1) = E'(d_{\mu-1})$. Therefore

$$\dim E(z) \cap E(b_\mu) = \dim E(z) \cap E'(d_{\mu-1}) \ge \mu \,.$$

Because $\mathfrak{y} \in (E(x) - E(z)) \cap E(b_\mu)$, we have $\dim E(x) \cap E(b_\mu) \ge \mu + 1$. Therefore $x \in S\langle v, \mathfrak{a}\rangle$. Because $z \in B$, we have $E(x) \cap E(b_0-1) = E(z) \cap E(b_0-1) = 0$. Hence $x \in Q$ by (A2). Now $f(x) = (g(x), h(x))$ is defined with $y = \mathbf{P}(\mathfrak{y})$ in $\ddot{E}(x) \cap \ddot{E}(a_0) = \{g(x)\}$ and with $E(z) = E(x) \cap W = E(h(x))$. We have $(y,z) = (g(x), h(x)) = f(x)$. The map f is surjective. Claim 4 is proved.

CLAIM 5. The map f restricts to a biholomorphic map

(A10) $$f : S^*<v, \mathfrak{a}> \to A \times S^*<w, \mathfrak{c}>$$

and $S^*<v, \mathfrak{a}>$ is biholomorphically equivalent to $\mathbb{C}^{\vec{\mathfrak{a}}}$.

Proof of Claim 5. Take $y \in S^*<w, \mathfrak{c}>$. Then $E(y) \leq W$. Now (A8) implies that $W \cap E(a_1) = E'(a_1 - 1) = E'(d_0 - 1)$. Hence

$$0 = \dim E(y) \cap E'(d_0-1) = \dim E(y) \cap E(a_1) \geq \dim E(y) \cap E(a_0-1) .$$

Therefore $y \in B \cap S<w, \mathfrak{c}> = C$. Hence $S^*<w, \mathfrak{c}> \subseteq C$.

Define $M = A \times S^*<w, \mathfrak{c}>$. Take $x \in S<v, \partial_j \mathfrak{a}> \cap Q$. Then $1 \leq j \leq p$ and $b_j - 1 \geq a_0$. By (A8) we have $W \cap E(b_j - 1) = E'(b_j - 2) = E'(d_{j-1} - 1)$. Therefore

$$\begin{aligned} \dim E(h(x)) \cap E'(d_{j-1}-1) &= \dim E(x) \cap W \cap E(b_j - 1) \\ &\geq \dim E(x) \cap E(b_j - 1) + \dim W - \dim V \geq (j-1) + 1 . \end{aligned}$$

Take $\mu \in \mathbf{Z}[0, p-1]$ with $\mu \neq j$. Then $b_{\mu+1} \geq a_0$ and (A8) implies that $E(b_{\mu+1}) \cap W = E'(b_{\mu+1} - 1) = E'(d_\mu)$. Therefore

$$\dim E(h(x)) \cap E'(d_\mu) = \dim E(x) \cap W \cap E(b_{\mu+1}) \geq \mu + 1 .$$

Hence $h(x) \in S<w, \partial_{j-1} \mathfrak{c}>$.

Take $x \in Q$ with $h(x) \in S<w, \partial_{j-1} \mathfrak{c}>$. Then $1 \leq j \leq p$. Also $0 \neq \mathfrak{y} \in E(a_0) - E(a_0 - 1)$ exists such that $g(x) = \mathbb{P}(\mathfrak{g})$. Then $\mathfrak{g} \notin W$. We have $\mathfrak{g} \notin E(h(x)) \cap E(b_j - 1)$ but $\mathfrak{g} \in E(x) \cap E(b_j - 1)$. Because $b_j - 1 \geq a_0$, we have $W \cap E(b_j - 1) = E'(b_j - 2) = E'(d_{j-1} - 1)$. Therefore

$$\begin{aligned} \dim E(x) \cap E(b_j - 1) &= 1 + \dim E(h(x)) \cap E(b_j - 1) \\ &= 1 + \dim E(h(x)) \cap E'(d_{j-1} - 1) \geq 1 + j . \end{aligned}$$

Take $\mu \in \mathbf{Z}[0,p]$ with $\mu \neq j$. Then $E(b_\mu) \cap W = E'(b_\mu - 1) = E'(d_{\mu-1})$ and

$$\dim E(h(x)) \cap E(b_\mu) = \dim E(h(x)) \cap E'(d_{\mu-1}) \geq \mu .$$

Also $\mathfrak{g} \in E(x) \cap E(b_\mu)$ but $\mathfrak{g} \notin E(h(x)) \cap E(b_\mu)$. Therefore

$$\dim E(x) \cap E(b_\mu) = 1 + \dim E(h(x)) \cap E(b_\mu) \geq \mu + 1$$

which implies that $x \in S\langle v, \partial_j \mathfrak{a}\rangle \cap Q$. Hence f maps $S\langle v, \partial_j \mathfrak{a}\rangle \cap Q$ bijectively onto $A \times C \cap S\langle w, \partial_{j-1} \mathfrak{c}\rangle$ for $j = 1, \cdots, p$. Taking complements implies that the restriction $f : S^*\langle v, \mathfrak{a}\rangle \to M$ is holomorphic and bijective. By assumption $S^*\langle w, \mathfrak{c}\rangle$ is biholomorphically equivalent to $\mathbb{C}^{\vec{\mathfrak{c}}}$. Also $A = \ddot{E}(a_0) - \ddot{E}(a_0 - 1)$ is biholomorphically equivalent to $\mathbb{C}^{a_0}$. Hence M is biholomorphically equivalent to $\mathbb{C}^{a_0} \times \mathbb{C}^{\vec{\mathfrak{c}}} = \mathbb{C}^{\vec{\mathfrak{a}}}$. In particular M is a manifold. Hence $f : S^*\langle v, \mathfrak{a}\rangle \to M$ is biholomorphic and $S^*\langle v, \mathfrak{a}\rangle$ is biholomorphically equivalent to $\mathbb{C}^{\vec{\mathfrak{a}}}$. Claim 5 is proved.

By induction, $S^*\langle w, \mathfrak{c}\rangle$ is dense in $S\langle w, \mathfrak{c}\rangle$. Hence M is dense in $A \times C$. Because $f : Q \to A \times C$ is holomorphic and bijective, $f^{-1} : A \times C \to Q$ is continuous. Hence $S^*\langle v, \mathfrak{a}\rangle$ is dense in Q. Since Q is dense in $S\langle v, \mathfrak{a}\rangle$, also $S^*\langle v, \mathfrak{a}\rangle$ is dense in $S\langle v, \mathfrak{a}\rangle$. Consequently, $S\langle v, \mathfrak{a}\rangle$ is irreducible and has dimension $\vec{\mathfrak{a}}$; q.e.d.

REFERENCES

[1] Andreotti, A., and Stoll, W.: Analytic and algebraic dependence of meromorphic functions. Lecture Notes in Mathematics *234*, Springer Verlag, Berlin-Heidelberg-New York, 1971, p. 390.

[2] Bloom, T., and Herrera, M.: De Rham cohomology of an analytic space. Invent. Math. *7* (1969), pp. 275-296.

[3] Bott, R., and Chern, S. S.: Hermitian vector bundles and the equidistribution of the zeroes of their holomorphic sections. Acta Math. *114* (1965), pp. 71-112.

[4] Chern, S. S.: Characteristic classes of hermitian manifolds. Annals of Math. *47* (1946), pp. 85-121.

[5] ——————: On the multiplication in the characteristic ring of a sphere bundle. Annals of Math. (2) *49* (1948), pp. 362-372.

[6] ——————: On the characteristic classes of complex sphere bundles and algebraic varieties. Amer. J. of Math. *75* (1953), pp. 565-597.

[7] ——————: Complex manifolds without potential theory. Van Nostrand. Mathematical Studies *#15*. Van Nostrand, Princeton, N. J., 1967, p. 92.

[8] Cowen, M.: Hermitian vector bundles and value distribution for Schubert cycles. Trans. Amer. Math. Soc. *180* (1973), pp. 189-228.

[9] Damon, J.: The Gysin homomorphism for flag bundles. Amer. J. of Math. *95* (1973), pp. 643-659.

[10] ——————: The Gysin homomorphism for flag bundles. Applications. Amer. J. of Math. *96* (1974), pp. 248-260.

[11] Ehresmann, C.: Sur le topologie de certains espaces homogénes. Ann. of Math. (2) *35* (1934), pp. 396-443.

[12] Ehresmann, C.: Sur le topologie de certain varities algébriques. C. R. Acad. Sci. Paris *196* (1933), pp. 152-154.

[13] Giambelli, G. Z.: Risoluzione del problema degli spazi secanti. Mem. R. Acc. Torino (2), *52* (1902), pp. 171-211.

[14] Hirschfelder, J.: The first main theorem of value distribution in several variables. Invent. Math. *8* (1969), pp. 1-33.

[15] ______________: On Wu's form of the first main theorem of value distribution in several variables. Proc. Amer. Math. Soc. *23* (1969), pp. 548-554.

[16] Hirzebruch, F.: Neue topologische Methoden in der algebraischen Geometric, 2. Aufl. Erg. d. Math. *9* Springer-Verlag. Berlin-Göttingen-Heidelberg, 1962, p. 181.

[17] Hodge, W. V. D.: The base for algebraic varieties of given dimension on a Grassmannian variety. Journ. London Math. Soc. *16* (1941), pp. 245-255.

[18] ______________: The intersection formulae for a Grassmannian variety. Journ. London Math. Soc. *17* (1942), pp. 48-64.

[19] King, J.: The currents defined by analytic varieties. Acta Math. *127* (1971), pp. 185-220.

[20] Lascoux, A., and Berger, M.: Variétés Kähleriennes compactes. Lecture Notes in Mathematics *154*, Springer-Verlag. Berlin-Heidelberg-New York, (1970), p. 23.

[21] Matsushima, Y.: On a problem of Stoll concerning a cohomology map from a flag manifold into a Grassmann manifold. Osaka J. Math. *13* (1976), pp. 231-269.

[22] Matsushima, Y., and Stoll, W.: Ample vector bundles on compact complex spaces. Rice Univ. Studies *59* (1973), pp. 71-107.

[23] Milnor, J. W., and Stasheff, J. D.: Characteristic classes. Annals of Math. Studies. Princeton University Press (1974), Princeton, N. J., p. 330.

[24] Pieri, M.: Sul problema degli spazi secanti. Rend. Ist. Lambardo (2), *26* (1893), pp. 534-546 and *27* (1894), pp. 258-273.

[25] Stoll, W.: A general first main theorem of value distribution. Acta Math. *118* (1967), pp. 111-191.

[26] Stoll, W.: About value distribution of holomorphic maps into projective space. Acta Math. *123* (1969), pp. 83-114.

[27] ———: Value distribution of holomorphic maps into compact complex manifolds. Lecture Notes in Mathematics *135*. Springer-Verlag. Berlin-Heidelberg-New York 1970, p. 267.

[28] ———: Fiber integration and some of its applications. Symp. on Sev. Comp. Var. Park City, Utah, 1970. Lecture Notes in Mathematics *184* (1971), pp. 109-120. Springer-Verlag. Berlin-Heidelberg-New York.

[29] ———: Value distribution of holomorphic maps. Sev. Compl. Var. I, Maryland 1970. Lecture Notes in Mathematics *155* (1970), pp. 165-190. Springer-Verlag. Berlin-Heidelberg-New York.

[30] ———: Deficit and Bezout estimates. Value Distribution Theory. Part B. (Ed. by R. O. Kujala and A. L. Vitter III.) Pure and Appl. Math. *25*. Marcell Dekker, New York 1973, p. 271.

[31] ———: Value distribution theory on parabolic spaces. Lecture Notes in Mathematics *600*. Springer-Verlag. Berlin-Heidelberg-New York, 1977, p. 216.

[32] ———: A Casorati-Weierstrass Theorem for Schubert zeroes of semi-ample holomorphic vector bundles (to appear).

[33] Tung, Ch.: The first main theorem on complex spaces, (1973 Notre Dame Thesis), p. 320 (to appear).

[34] Vesentini, M. E.: Construction géométrique des classes de Chern de quelques variétés de Grassmann complexes colloque de topologie algébrique, Louvain, 1956, pp. 97-120. Georges Thone, Liege Masson & Cie, Paris, 1957.

[35] Wu, H.: Remarks on the first main theorem of equidistribution theory I - Journ. of Diff. Geom 2 (1968), pp. 197-202, II - *Ibid.* 2 (1968), pp. 369-384, III - *Ibid.* 3 (1969), pp. 83-94, IV - *Ibid.* (1969), pp. 433-446.

INDEX

ample, 40
amplification, 40

basic Chern forms, 38, 39

Chern form, 36
  basic, 38, 39
  of symbol α, 59
  total, 36
classification sequence, 39
compensation function, 8
complex projective space, 11
connection matrix, 35
counting function, 8, 9

deficit, 8
de Rham groups, 38
double intersection diagram, 64
double representation, 71

First Main Theorem, 8
flag space, 16
  short, 17

general linear group, 12
general position, 67, 68
Grassmann cone, 11
Grassmann manifold, 11

holomorphic frame, 35

independent, 12

locally trivial, 43

open Schubert cell, 30

p-plane, 12
projective map, 43, 44
pseudoconvex exhaustion, 6

representation, 67, 69

Schubert cell, open, 30
Schubert family, 27
Schubert variety, 28
short flag space, 17
symbol, 15

tautological bundle, 19
total Chern form, 36
triple intersection diagram, 83

unitary group, 14

valence function, 8

Library of Congress Cataloging in Publication Data

Stoll, Wilhelm.
Invariant forms on Grassmann manifolds.

(Annals of mathematics studies ; no. 89)
Bibliography: p.
Includes index.
1. Grassmann manifolds. 2. Differential forms.
3. Invariants. I. Title. II. Series.
QA331.S859 514'.224 77-85946
ISBN 0-691-08198-0
ISBN 0-691-08199-9 pbk.